Toy Bop

KID CLASSICS
of the 50's & 60's

By Tom Frey

PHOTOGRAPHY
Jim Douglas

GRAPHIC DESIGN
Dick Brodeur

FOREWORD BY
Ira H. Gallen

To Beth & Ben
Have fun
in
Toyland!
Tom Frey
"2016"

DEDICATION

This book is dedicated to
Laurel, Randy, and Julie Frey

TOY BOP
Kid Classics of the 50's & 60's
by Tom Frey

Photography by Jim Douglas

Graphic Design by Dick Brodeur

Foreword by Ira H. Gallen

Published by
FUZZY DICE PRODUCTIONS INC.
4556 WM. PENN HWY., SUITE 150
MURRYSVILLE, PA 15668 USA

Library of Congress Catalog
Card Number 93-074676
ISBN 0-9639700-0-3

Color Separations by Scantrans Pte. Ltd.
Printed in Singapore
by Toppan Printing Co.(S)

First Edition

TABLE OF CONTENTS

FOREWORD

*S*everal years ago, I was invited by the Smithsonian Institution to do a special video presentation sharing my collection of vintage television commercials with an audience of members from the museum. Their age group varied somewhat but were well represented by former Baby Boomers. Since both they and television as a medium shared the same era of infancy, this subject matter was considered by the Smithsonian and everyone seated to be part of America's pop cultural heritage. While they watched, laughed, and eventually sang along to commercial jingle lyrics, lyrics locked up in their heads and consciously forgotten for decades, I watched something different. I saw a room full of strangers unite with a common bond. Having grown up together under the influence of television's hammerhead hype, we somehow hold, in fond memory, sponsors telling us what to eat, which sneakers were the fastest, and what toys we should be the first in the neighborhood to buy. Of all these images from TV land, the ones with the biggest reactions and applause were the toys.

Today we as a generation have reached an age triggering the need to look back. Let me introduce to either those outside the toy collecting hobby or others unacquainted, someone who not only understands all this but lives it with a passion, Mr.

Toys himself, Tom Frey. Tom first entered the arena of collectible toys over a decade ago setting out on a typical mission to recapture his childhood favorites. In the process, he was immediately taken over by the uniqueness and charm of these toys beyond his own narrow neighborhood experience. He said they seemed to cry out for their story to be told, a story reflecting the post-war Baby Boom era, the advent of television, and specifically, the toy industry's imaginiative achievements created while catering to an unprecedented mass market. As a result, Tom became one of playtime's pioneers with articles in the hobby's leading publications telling tales of toys gone by as only he can.

Now may I welcome you to Toy Bop, a playroom of the past. It is an astounding assemblage of our toy faves, a wishbook of all those classic years rolled into one. Opening its cover is like raising the lid to a giant toy chest containing a Who's Who or more appropriately a What's What assortment of the playthings we had and others we so desperately wanted. So return with us for a nostalgic twinge. Slip your Hoppy sixguns back in their holster and set Betsy Wetsy down in her crib. The cookies and milk are on the counter. I'll be in the other room if you need me. It's Howdy Doody Time!

Ira H. Gallen

As Seen On TV

In the 50's, society was undergoing a transformation. For the first time television was beaming into the homes of over 50 million people. Commercials tempted and teased with everything from the latest household miracle products to cookies, candy, cereal, and something kids found most tantalizing ... Toys!

The honor of being the first toy commercial ever broadcast went to Hasbro's Mr. Potato Head whose captivating performance opened a year-round window to Toyland previously unrealized. Other companies soon followed until by the end of the decade nearly all the major American toy manufacturers had entered the small screen. All but one.

On the cover of *Time* magazine December 12, 1955, toy tycoon Louis Marx was pictured in the foreground of Santa Claus with toys floating overhead. In the article, Marx took pride in his miniscule advertising budget (an annual amount of $312). Up to that point, his company had resisted promotion with the newfangled medium of television. However, by 1959, his attitude reversed as Marx finally decided to go after the TV market in grand style. In a retailer's promotional brochure his plan of attack to bombard an estimated 27 million kids with a television ad campaign of strategically placed toy commercials was unveiled. Exposure to his three-month holiday season blitz was predicted to tally one billion, 499 million viewings. Then like Hasbro's *hot potato*, Marx found testing the waters of television to be a warm reception and assigned his logo stickman, Magic Marxie, to extended duty in TV Land. Now to fully appreciate the combined promotional push, expand this volume of commercials with broad-casts by Hasbro, Ideal, Remco, Mattel, and others. Soon the significance of the words "As seen on TV" comes into focus.

Meanwhile on the home front, kids busy twisting rabbit ears and attaching crumpled balls of aluminum foil to television antennae beheld shimmering images of each company's wares. Before their very eyes, toys performed cleverly choreographed play to the tune of melodic jingles. Throughout all this enviable excitement, the tiny tots sat helplessly mesmerized. Little arms raised toward the flickering gray screen pointing to the

commercial's object of delight. Then with squeaky voices aimed at the kitchen, came the inevitable plea, "I WANT THAT TOY!"

This battle cry echoed across the country as the original television generation lobbied practical parents for toys "As seen on TV!" So strong were those "words from our sponsor" that video veterans remain haunted to this day with the heavy hype from decades past. Slogans still bouncing along brain waves include "Every boy wants a Remco toy, and so do girls!" or "It's a Wonderful Toy, it's Ideal!" the bird squawking "It's Kenner! ... It's Fun!" and 60's lingo "You can tell it's Mattel ... It's Swell!"

How swell? How wonderful? How fun? ... Let's just say if you were the lucky recipient unfolding box flaps to unveil one of those coveted classics, what often came out were gigantic plastic playthings exploding in color and ready to roll. Kids took to the controls in battery-operated bliss while experiencing their first objects of owner-ship ... experiencing what many believe to be (dare I say) the best toys ever made.

Since then, the toys of yesteryear have actually developed another function not listed on the instruction page–that is, the power of time travel. Coming face to face with old favorites can flood the mind with images of family and friends, places and events gone by. The hand is once again raised, irresistibly pointing, "I HAD THAT TOY!"

The Attic That Time Forgot

For some, a mere trip home to the attic may discover a parked fire engine, Dolly and her wardrobe, or a set of six guns waiting in folded holsters for their little sheriff's return. For others, the trigger to the past doesn't come as easy. Many were the parents who tucked away baseball gloves, a fish bowl, a pair of training wheels, or hand prints petrified in clay. These mementos are all very fine setting the mood for a childhood toy search but "Where's my Robot Commando? What happened to Mr. Machine and Mighty Matilda?" Where are those big plastic playthings now famed as the American Toy Industry's greatest hits from decades past?

Guess what? They're back! Toys long believed extinct are beeping and creeping around the countryside. Bizarre creations like Big Loo, Great Garloo, King Zor, and Odd Ogg are being summoned by their old masters. Yes, the toys that made the fifties nifty and the sixties super are appearing at toy shows and flea markets near you. And this time you don't even need Mom's permission to play!

Re-Stocking The Toy Chest

Suppose you are able to travel back in time solely for the purpose of a shopping spree into an old toy store. Select any year to purchase crispy-clean, factory-sealed playthings at their original prices! There are only two restrictions. First, you only get one trip. Second, whatever comes back has to fit, with you, in a time machine the size of a telephone booth. Ponder this fantasy awhile and we'll start over.

Okay, snap out of it. Forget the time machine. I heard it uses a gallon of propellant per year traveled back with fuel costs of $1,000 per gallon. More importantly, what I secretly did for some of you, was prioritize a wish list. Now instead of returning to an old store, we enter one of today's super-sized toy shows. It's like a scavenger hunt into the past for a hands-on look into childhood. Now, chances are, one or more of the selections you were about to stuff into the Time Booth are sitting here on tables. Some may even be displayed in original packaging undisturbed since leaving the factory decades ago. Best of all, differing from a museum exhibit, each and every one is for sale!

Too good to be true? Now for the fine print at the bottom.

First, as one might guess, the prices have gone up. In the early 60's most top-of-the-line toys retailed around $10. Today that same toy may equal the price of a fill-up at the gas station, a week's groceries, your monthly car payment, or even a mortgage installment. This pricing variation depends on the demand for that particular item, its condition, completeness, and of course, who's selling it.

It is most important that a collector know exactly what it is they are buying. Even memory of a favorite toy gets fuzzy after several decades. Don't accept "It's all there" or "It worked when I last tried it" until you've checked for yourself. If the box and directions are present, further questions may be answered. Actually, the best sources I've found to brush up on a toy's inclusions or functions are vintage department store Christmas catalogs. Typically they show each item, often in use, with accessories and a somewhat embellished description. Underneath the listing is a catalog code number and original selling price everyone wishes was still a phone call away. Unfortunately, few of these volumes have survived the years and when found can be as pricey as the toys themselves. Further information can be discovered in the hobby's books, magazines, and videos. Devour whatever means of study available. Why? Because it's as important to know what and when to buy as what and when not to buy.

Now, lace up those sneakers, gulp some caffeine, and stay sharp. The doors are about to open to Toyland. You're ready. You've trained well for this event. Every toy related family photo has been scrutinized for details. A tested set of alkalines bulge from your pocket. A wad of "toy money" has either been saved up or more likely "borrowed" from family funds. All that's left is to beat out scores of other avid collectors looking for the same toys. To make matters worse, many dealers are themselves collectors who snare the exquisite or scarce *objet dé fun* prior to the public's entry. Hey, no one said this was going to be easy.

Finally, one last problem ... availability. What I'm about to say may shock you with a reality some refuse to believe. That is, some mass produced toys, twenty or more years old, may actually be out of existence. Let's further clarify this statement by referring only to those found in the condition collectors seek, mint in the box. Again for the purpose of current market availability, I also disqualify examples locked up in museum or personal collections till kingdom come.

How can this be? Wasn't any one person in charge of preserving these great toys while we were out learning the skills of our trade? ... The answer ... no (although I'm sure my wife disagrees). We must realize that the intended purpose of a toy is to be played with by a child. Getting used and abused, few toys have survived the wrath of these destructive cherubs. Other once delightful playthings simply outlive their usefulness and eventually disappear having been betrayed by our disposable throwaway society. Yet somehow, a chosen few remain. Unfortunately, of those having slipped through the crevices of time, most are likely to show the scars of playtime past.

Collectors seem to forget that these toys were never meant to be used thirty or forty years later. In the era of their production, they were simply ... toys. Their design and composition did not intend for them to be carried into the future as collectible artifacts. Often, those saved by parents in attics and basements or even unpurchased store stock haven't aged well. Many plastic toys warp, fade, discolor, and become brittle. Metal toys scratch, dent, and rust. Battery motors are subject to seizing with dry-rotted rubber drive belts or corroded contacts. Add to this list, inherent flaws of the manufacturing process or play-wear and it becomes evident that a serious collector can expect an almost endless task of upgrading.

Frankly, I was elated the day I discovered toy collecting just plain existed! Not only did it exist but was a thriving international hobby. Immediately my mind assembled a top ten list of favorites I was sure could never be found. But, as the story goes, they were. They came back mostly due to perseverance and, better yet, luck. To those who are about to embark on the exciting mission of toy collecting, I wish that luck.

At this moment, a van full of your long lost favorites is heading for the next toy show. An adventure awaits.

THE TOY HALL OF FAME

Why not? No one ever flinches at the acclaim given to athletes who throughout time have done something stupendous with a ball. Now let's hear it for the inventor of the ball itself! Or the ingenious men and women who dreamed up Teddy bears and Barbie dolls, Mr. Potato Head and Mr. Machine, Erector Sets and electric trains. Yes Virginia, there is a Toy Hall of Fame. The Toy Manufacturers of America (TMA) annually inducts those individuals whose lifetime efforts have had some Santa Clausian effect on the industry. Toy moguls like Louis Marx (Marx Toys), A.C. Gilbert (A.C. Gilbert Co.), Joshua Lionel Cohen (Lionel Corp.), Herman Fisher (Fisher-Price), Ruth and Elliot Handler (Mattel Toys), and Benjamin Michtom (Ideal Toys) are among those merited

for lifetime achievement by the Toy Manufacturers of America. Some others having a dramatic effect on the industry without actually being a toy manufacturer themselves include Walt Disney, Jim Henson, and inventor Marvin Glass.

With the vast list of characters coming from both Walt Disney and Jim Henson's productions, it's easy to see their impact on the toy industry. However, the other aforementioned individual, Marvin Glass, may not be a household name as Disney and Henson, but without whose achievements, this book would be reduced to pamphlet proportion. Mr. Marvin Glass was a toy inventor extraordinaire. This man's life story is so astounding the nearest comparison would be a cross between Willy Wonka and Howard Hughes.

In *A Toy Is Born*, Marvin Kaye gives an account of daily operations at the inventor's facility. Kaye descibes the peculiar working conditions self-imposed by a company in constant fear of industrial espionage. Taking steps to thwart competitor spies, Glass and his associates were holed-up in a windowless Chicago factory with armed guards posted at the entrance and exit points. Inside, secrets of new toys in development were further protected by segmenting the parts and mechanisms in different locked departments with only Mr. Glass himself able to see and understand the projects in their entirety.

The unmistakable style thriving in that fortress of fun was functionally unique, always colorful, and most often, immense. A sampling of 60's toy greats from the drafting tables of Marvin Glass and Associates include Ideal's Mr. Machine, Robot Commando, King Zor, Odd Ogg, Smarty Bird, Gaylord, and the Mouse Trap Game. Other company's notables included Marx Toy's Rock'em Sock'em Robots, Irwin's Dandy the Lion, Hubley's Golferino, and Eldon's Yakkity Yob. I simply find it impossible for anyone to comprise a list of favorite 60's toys not highlighted by a Marvin Glass design. As a trademark of their work, a small black and white MG logo was printed on toy boxes to identify Marvin Glass as the toy's inventor. Former Vice President of Irwin Corp. Bertram Cohen recalls their spectacular Dandy the Lion of 1963: "There was excitement at Irwin as we acquired our first Marvin Glass toy. You may notice that we stamped the underside of Dandy with the inventor's company name, Marvin Glass and Associates."

We who celebrate the excitement and wonder of youth through childhood artifacts are greatly indebted to this man and his talented troupe for grabbing the toy market by the ankles and smacking new life into its consciousness. For if the founders of the American toy companies can be likened to the role of Santa Claus, then truly Marvin Glass and Associates were the elves.

Now rather than follow an inventor like Glass, I'd like to precede his career with the man who once saved Christmas and the toy market itself, Toy Hall of Famer, A.C. Gilbert. Although Mr. Gilbert's company was reknowned for a diverse range of toys from Mysto-Magic kits to chemistry sets, American Flyer trains, and eventually secret agent spy toys, his biggest hit with America's construction minded and engineers of tomorrow, was undoubtedly the Erector Set. Creation of this classic alone would be just cause for induction among the country's other toy titans, but I would like to relay a story found appropriately from *Inside Santa's Workshop* by Levy and Weingartner. The incident took place in 1917.

That was the year Woodrow Wilson had entered America into World War I. Amidst the anxiety and paranoia caused by war, the US Council for National Defense was proposing an embargo on the sale of Christmas gifts so as to divert labor and materials toward the war effort. Though a well intentioned plan, implementing such an embargo would likely do more harm than good. It would certainly cripple the toy market and stunt the economy along with it. Many were outraged but helpless to defy this group of starched collar stiffs. Was all hope lost? Who could find a way to stop this toy tyranny? Who would even have authority to present an opposing viewpoint to the president?

Enter, A.C. Gilbert. At the time Mr. Gilbert was himself a president in two capacities. First of

his own toy company and secondly as the official so nominated to represent the Toy Manufacturers of America. Gilbert quickly did his homework researching the amount and types of materials used or anticipated by the War Department. Gathering his findings, he also packed a sack of toys and headed for Washington. A somewhat convincing speech was delivered explaining to the committee how detrimental their proposal would be to business, the morale of society, and especially America's youth. His statistical findings provided credible evidence that toy materials were either not being used for military needs or were of an insignificant amount industrywide. The clincher closing all arguments was the act of Gilbert dispersing various playthings to the council for close examination. Soon there was a room full of grown men playing with toys. Reportedly, the Secretary of the Navy latched hold of an Ives submarine and wouldn't let go. A motion was passed allowing the toy industry to continue without intervention.

At this point I can only imagine the gavel crashing down to ignite a flurry of reporters seeking statements for the front page news. Then after a series of camera flashes flicker and fade, A.C. Gilbert is free to gather up his empty sack and quietly exit from the hustle and bustle of the nation's capital. Pausing for a moment outside he looks skyward and it's as if the sparkle in a million kid's eyes light the way ... Thank you, Mr. Gilbert, and may all your Christmases be white.

YOU ARE THE MASTER !

"Mine at last!" cried the evil sorcerer finally grasping Aladdin's lamp with a greedy grip. Now there would be no end to the power and riches he longed for.

I'm not suggesting that kids identify with the villain but rather liken themselves to his pre-lamp syndrome, being powerless and penniless. Exceptions aside, most kids grow up in this condition totally dependent upon parents for their every want and need. Relating to toys is, first, the obvious. Kids must sharpen skills lobbying Mom or Dad for the neat novelties they saw on TV, in a store, Wish Book, or at a friend's house. Usually those pleads and prods are most effective approaching a birthday or holiday season. Other times they fall on deaf ears. As for piggy bank finances ... they just do not, a great toy, buy.

So much for the magic genie's riches. What about power? As kids, not only didn't we have a chance to win the hand of the fair princess, but everything said and done required ... permission! Our lives were not our own where, without a moment's notice, we were piled in the car for an instant field trip. It was time to observe everyday American society in supermarkets, shopping centers, to have our heads barbered, or to visit polka-dotted aunts who pinched cheeks and on occasion might exhale smoke from their nose like a dragon!

Finally to rescue us from our punky plight came ... the toys! Not just any toys. No decaffeinated distractions like building blocks or paint by number sets. I mean power toys as in "Oh do my bidding thou toy of the titans." I mean grabbing the voice-controlled wired remote of Robot Commando and instructing him to fire rockets and pitch missile balls. I mean ramming a cannon ball or two down the barrel of the Johnny Reb Cannon, and boinging them across the room. Launch fighters, missiles, and torpedoes from the deck of Battlewagon, or satellites from Project Yankee Doodle. Even Lionel Trains promised kids the feeling of control with commercials boldly boasting "You're the Boss! with a Lionel train." The very words "You are his Master!" were inscribed on the box encasing Marx's Great Garloo. Garloo, a worthy ally, was a two-foot-tall green scaly sea monster and, most importantly, yours to command!

Okay, so we got a little carried away with the soaring projectiles. And it's not as if holstering up cap guns would save you from having to eat asparagus or clean your room. Even King Zor couldn't get me out of attending a girl cousin's roller skating birthday party. But what these secret weapons really did for us was unleash our spirit of fun, like the so called "magic feather" which gave Dumbo the confidence to fly, an ability he had inside all along. And fly we did.

ONCE UPON A TOY

Know Ye this! For every Frisbee, every Silly Putty, Slinky, or Mr. Potato Head there are a hundred other phenomenal ideas yet undiscovered right under your eye-browed nose glasses. Toy origins have been as varied as the products themselves. Once developed, their concepts often seem so simple and obvious it teases and taunts those of us who looked the other way while others with calculated talent or dumb luck cashed in on crazes and fads. From Marvin Kaye's *A Toy Is Born*, which as the title suggests reveals some of their humble beginnings, I learned of the world's first "Teddy Bear."

One day in 1902 President Theodore "Teddy" Roosevelt and his hunting buddies took the day off for adventure in search of big game. The trophy sized wildlife must have read about the upcoming event and headed for the hills leaving behind a baby bear cub. Eventually, the hunting party came across the animal, captured it, and offered honors of "the kill" to Roosevelt. When the President refused to shoot and demanded the baby cub be set free, the story took the nation's fancy. This gave Brooklyn toy store owner Morris Michtom the idea to make a brown plush bear which he placed in his window with the sign, "Teddy's Bear." Prior to doing so, Michtom had written to the president asking for permission to use his name in association with the stuffed animal. The request came as a surprise to Roosevelt who was doubtful that his name would have any effect on selling toys, however, permission was granted. Since then, everyone has grown up dragging around well loved Teddy Bears. Yet few people are aware of

how the stuffed bear acquired the name "Teddy." Fewer still are aware of toy store owner, Morris Michtom, having such success with his Teddy Bears, that the proceeds enabled him to found the Ideal Toy Corporation.

Stories like this abound throughout Toyland like the icing dripping from Hansel and Gretel's cottage. It is believed that A.C. Gilbert got the idea for Erector sets while riding over a similarly constructed bridge. Herbert Schaper who hobbied the carving of wooden fishing lures one day fashioned a bug-like creature that became Cootie. Silly Putty was the result of chemist James Wright's failed attempt to revolutionize rubber while the idea for Slinky came about as Richard James noticed coiled scrap cut-aways bouncing from the floor as they fell from machinery. Meanwhile, housewife Ruth Handler discovered her daughter Barbie was much more interested dressing teenage paper dolls in fashion clothes than changing diapers with three dimensional infant dollies. A void in the market place was realized. Although her son Ken didn't share the enthusiasm, hubby Elliot did which inspired the Mom and Pop entre-preneurs to begin their Mattel Toy company. And finally, even name calling in a domestic squabble was put to creative use by workaholic toy inventor Marvin Glass who turned an intended slur into the toy sensation, Mr. Machine.

So start dreaming and scheming. All it takes is a thought. Then another. Roll them together into an idea and voilá, you're moving in on Easy Street. Hey that sounds like a good name for a game!

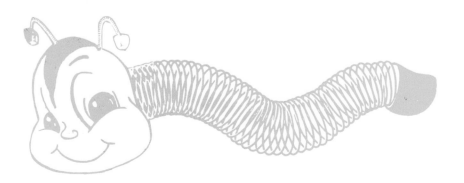

Oh Toy My Toy

I'm still wondering who filled all the occupations in life other than Cowboy, Astronaut, Fireman, Ballerina, Schoolteacher, and Nurse. Of course, no kid ever had to grow up to fulfill their dream careers. The American Toy Industry saw to it that yesterday's youngsters were well equipped.

The 1950's and 60's first saw flying saucers called Frisbee tossed overhead to land in toy history. Kiddie Cape Canaverals launched rockets inspired by NASA's space race with Russia. When playtime re-staged one of the countless westerns from television, kid pokes holstered up replicas of their favorite hero's shootin' irons to blast caps till high noon. Meanwhile across the street, backyard maneuvers were conducted with olive drab toylines that went Rat-a-tat-tat. And where were the girls? Why they were busy baking cookies in light bulb ovens while chatting with dollies that walked, talked, drank from bottles, wet their pants, and cried real tears.

Finally, after all this serious play was taken care of, we bounced some Super balls, twirled Whirly Twirlys, got silly with putty, monsterized magnets, and blew bubbles of Super Elastic Bubble Plastic. And yes, as legend goes, everyone wiggled their hips trying to hula with a hoop.

What follows are photos and descriptions of Grand Poobahs from that "not so long ago." The classic toys of this era represent an age of opulent consumerism running wild in its formative years. Although once mass produced, highly promoted, and plentiful; relatively few exist today. Those that remain stand as a reflection of the art and design of an earlier time. As today's industry strives to improve child safety, limit breakage, and minimize costs, many of the features these toys were best known for have been watered down or eliminated.

As society has changed so has toy society. Now for my friends who joined me at play and for those I've yet to meet, may I present ... our toys.

*As the years go by, the little hands which once fumbled to load a toy with batteries
have grown and are now carrying hard hats and attaché cases.
Faded memories of fun are buried under piles of maturity, responsibility, and ambition.
Yet, somehow, the short period of time known as childhood
holds a lifelong fascination.*

Long before the first word is uttered, before focused vision or standing erect, innocent newborns lying in bed find themselves in an avalanche of crib toys. Teddy bears headline the list of stuffed animals. Rattles are often close at hand. Musical mobiles dance overhead while a busy box gets frequent demonstrations by onlookers. Regardless of the child's gender, you can bet Uncle Ned will toss in a football.

Many of the toys from early childhood are long gone and forgotten. Yet others somehow defy the odds of "age recollection" and are remembered anyway. Check your family archives. Blow the dust off photo albums and drag out that clunky projector. A few images may remain on black and white jagged edged photographs or eight millimeter muted performances of family and friends squinting into the blinding spotlights. Wait for the dots in your eyes to fade and then let's peek into the past.

THE BIG PARADE
Marx 1964

No portrayal of Toyland would be complete without toy soldiers in parade dress. They stand proud as a symbol of the fun to come. The Big Parade lives up to this tradition by following the drummer's beat with an arm swinging leg stepping march. To steer the team, simply adjust the flag to turn the center wheel. Then join in the merriment with a miniature baton provided for kids to personally lead the parade.

HUMPTY DUMPTY TAKE APART TOY
Playmaker Toys 1950's

Have your great fall and then put Humpty back together again. In case you forget the rhyme, it's printed on the back of our animated friend's egghead. Kusan also produced a yellow faced version with green clothing.

BUNNY BANKS
Knickerbocker 1950's

It appears as if Mr. and Mrs. Bunny hippity hopped in from a Max Fleischer cartoon. What I don't get is ... with all the carrots they must eat, why the eyeglasses?

POPEYE THE WEATHERMAN
Colorforms 1959

Early sets like Popeye the Weatherman provided solid color play pieces with simple black line detail to stick-on and remove from the scene board. The theme of this set was to teach kids how to dress for the weather and was used as an activity on the *Romper Room* television show.

OLD MAID

Tillie
Tumble

Husky
Hank

Lotta
Noise

Hedda
Howell

Hiram
Hay

Betty
Bumps

Wacky
Witch

Hugo
Hunt

Dippy
Dabb

Kuku
Klown

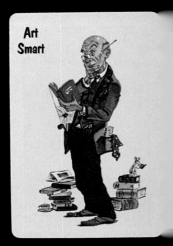

Art
Smart

John
Law

Fanny
Flint

Gusto
Graft

Molly
Moo

NORVAL THE BASHFUL BLINKET
Revell 1959

What's new in Dr. Suess' Zoo? How about taking apart Norval the Bashful Blinket to see how many wacky ways there are to re-assemble him. After exhausting every possibility try the same trick with his friends Gowdy the Dowdy Grackle and Tingo the Noodle Topped Stroodle.

Norval the Bashful Blinket appears courtesy of Dick Brodeur

FIREBIRD 99 DASH
Remco 1958

This is where it all begins: parent puts child behind the wheel. Most often before infants can speak or hold their own spoons, they find themselves turning the keys to their future behind a kiddie car dash. Although Remco's stunning console drove kids "steer crazy" they also beeped the horn, lit turn signals, and constantly had the windshield wiper wagging back and forth. A driver's license with a road map on the reverse goes in the glove compartment which is where we put everything else except, of course, gloves.

THE THREE STOOGES HANDPUPPETS
Ideal 1960's

Chowder Heads Moe, Larry, and Curly brought their pie battles, violin plucking eye pokes, and bass drum belly bops to TV in the late 50's immediately turning kiddie audiences into "victims of coicumstance." These hand puppets were favored among other 60's toys available for kids to stooge around.

WHAT'S UP DOC?

"Calling Doctor Howard, Doctor Fine, Doctor Howard!" "Calling Dr. Kildare." "Calling Ben Casey." Every kid became doctor for a day wearing those nose pinching eyeglasses while testing out an array of plastic instruments on somewhat reluctant pets, family, and friends. Of course, Junior Doctor waiting rooms were known to empty the minute candy medicine ran out.

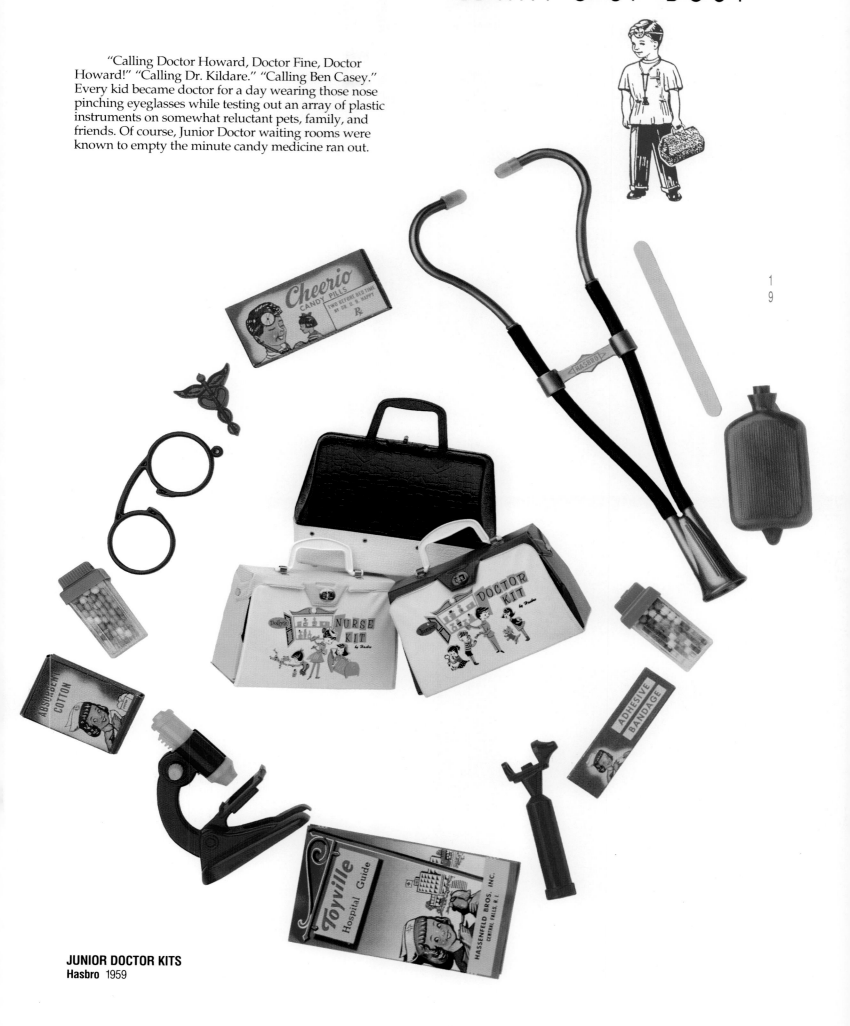

JUNIOR DOCTOR KITS
Hasbro 1959

Amaze your friends! Be the life of the party! Just three taps of the magic wand and Presto! the handkerchief changes color, or the coin disappears, the flower squirts, or the Queen of Hearts comes down dancing. Whatever the trick, I can neither figure out nor perform without dropping something. Maybe I should have paid closer attention to a certain magical moose whose opening line goes, "Hey Rocky, watch me pull a rabbit out of my hat!"

FROGGY THE GREMLIN
Rempl 1950's

This mischievous friend, famous for popping up in a puff of smoke, first appeared on television in 1954 with Ed McConnell of *Smilin' Ed's Gang*, then in 1957 with Andy Devine of *Andy's Gang*. But first you must say the words, "Plunk your magic Twanger Froggy!"

SNEAKY PETE'S MAGIC SHOW
Remco 1958

This is the magic toy which featured the famous Cut the Lady in Half trick. Sneaky Pete's Lady and her "bed of nails" were included in each of three sets including the smallest, Broadway, the medium sized, Professional, and the deluxe, Master set. Other hocus pocus highlights up Sneaky's sleeve were the Mystic Vanishing Box, Oriental Mystery Chamber, and the crowd pleasing Magic Flower Pots.

MAGIC BUNNY
Irwin 1950's

The "magic" involves pressing Bunny's tail for balls to disappear from his hat and pop out of his mouth with "squeaks of delight."

RED RAVEN MOVIE RECORDS
Morgan Development 1956

This is incredible. At one time or another, everyone has picked up flip pages which demonstrate the technique of cartoon animation. A quick thumbing of these pages seem to bring the progression of drawings to life. Basically, Red Raven movie records perform this same display with a Magic Mirror top resembling a carousel. A series of 78 rpm records colorfully set the animated scenes into motion which relate to song themes being played.

Initially the records were made of paperboard with a metal band supporting the perimeter. These were six inches in diameter with the record grooves directly over the graphics. Later these were replaced with solid color eight inch vinyl records which displayed the movie from paper labels. Both the orchestra and singing are wholesome and hokey, the sound effects ... astounding!

POP GOES THE WEASEL

2
2

Even though these were made for toddlers, didn't it seem as if older brothers and sisters did most of the weasel popping? And why is it that regardless of the character inside, we commonly refer to these as "Jack-in-the-boxes?" So what should it be, pop-out character music boxes? The moving eyes clown version shown was popular throughout the 60's and can be spotted in the Treasure House on early episodes of *Captain Kangaroo*. Mattel advertised other character Jack-in-the-boxes on their *Matty's Funday Funnies* television show resulting in Matty being pictured on a number of their lids. Since these were originally sold in printed cardboard boxes, I suppose the collector's ultimate would be to find a mint-in-the-box Jack-in-the-box ... right?

Another Mattel musical toy with the same concept is a scrumptious tin litho pie which cranks the tune *Sing A Song Of Sixpence*. At the conclusion, out pop five blackbirds who are then re-inserted by pushing down the cherry.

Today when re-discovering these toys, the surprise isn't really who pops out, but instead, after all these years that many can still carry a tune.

BUGS BUNNY IN THE MUSIC BOX
Mattel 1962

CASPER IN THE MUSIC BOX
Mattel 1959

MUSICAL BLACKBIRD PIE
Mattel 1953

FLIPPER IN THE MUSIC BOX
Mattel 1966

POPEYE IN THE MUSIC BOX
Mattel 1951

EVERYONE KNOWS IT'S SLINKY

Something went Boing! As legend has it, in 1943 Richard James noticed a coiled spring bouncing from machinery at a Navy Shipyard and saw the possibility of developing it into a child's toy. After some experimenting with various gauge and diameter coils to find what seemed the right combination, Richard turned to his wife Betty and her dictionary for naming their new novelty. Descriptively, the slinking motion seemed most appropriate so, "How about Slinky?"

Today, the super successful slinking sensation continues to travel down our stairs and through our heads with the timeless jingle, "It's Slinky. It's Slinky, for fun it's a wonderful toy. It's Slinky. It's Slinky, it's fun for a girl or boy." Besides the hand-held stair slinkers, these Slinky animals and vehicles stretched the imagination along with the spring. Each are made by James Industries and date from the 1960's.

SLINKY DOG   **CATER-PULLER**

PAT AND MIKE'S HAND CAR

SWAMP BUGGY

Ever since the first baby doll squeaked "Mama", kids have been clamoring for talking toys. In the 60's Mattel devised a pullstring device to play a record with eleven or so random sayings which revolutionized and dominated the market. Mattel's new mechanism provided excellent sound clarity, volume, and reliability vastly improved in comparison to previous hand crank talking toys. Many licensed cartoon and TV characters were produced in this series speaking recorded phrases of celebrity voices. Along with the special attention given sound quality, Mattel's new dolls and hand puppets were exquisitely made of colorful plush and cloth bodies with three dimensional vinyl heads. If you don't see some of your favorite talkers in this chapter, stay tuned. I'll see if I can "pull strings" to have others appear with other related themes.

LINUS THE LIONHEARTED
Mattel 1964

Linus came a long way from his early cereal commercial days avoiding stampedes by *Crispy Critters*. Eventually Linus really did become king of the jungle in his own TV show with much credit going to his spokesman, Sheldon Leonard. Some of our cartoon king's proclamations were "I'm a work of art with a Lion's heart" and "Sorry, this Li-on's busy."

BOZO THE CLOWN
Mattel 1963

"Howdy, I'm your old pal Bozo!" greeted a decade full of kids who played with the world's most famous clown. Bozo's very name is even synonomous with the word clown. Much of this adieu is due to Larry Harmon's voice characterization being second to none as our Big Top buddy's creator. Mattel's talking Bozo brought our friend to life with indisputable statements like "Yowl Wee Ka-Zowie," "That's a real rootin' tootin' trick," and his advice to generations of friends, "Just keep laughing!"

BUGS BUNNY
Mattel 1961

Possibly the single most popular animated talking doll, Bugs has evolved through at least six design variations and counting. Some favorite quotes include his trademark "What's up Doc?" also "Hey-ah take me with ya," and "Now let's take it e-e-e-easy." It's interesting to see that this particular talking toy survives a nuclear war in the film *Mad Max* ... and still talks!

MATTY MATTEL
Mattel 1961

Mattel Toys' crowned king of kid-dom, Matty Mattel takes a break as spokesboy offering playtime conversation like "Let's have a picnic," "Let's play cowboy," or "I like to play with you."

SISTER BELLE
Mattel 1961

The girl half of Mattel's famous brother and sister act, Belle, makes friends with sugar coated sweet talk like "I think you're nice," "Let's draw pictures," or "Sing me a song." She repeatedly told playmates "I'm glad we're friends" and they were too.

MATTEL, INC. TOYMAKERS

YOU CAN TELL IT'S MATTEL IT'S SWELL !

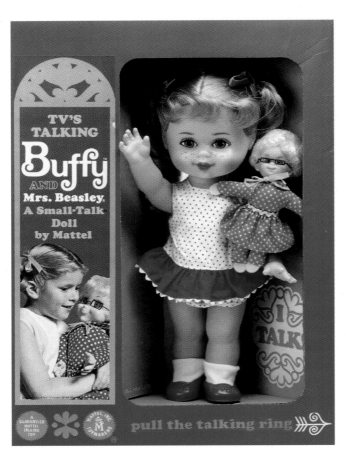

BUFFY AND MRS. BEASLEY
Mattel 1968

Everyone adored Buffy on *Family Affair* and recall her being inseparable from the Mrs. Beasley doll. This pint sized re-creation of Anissa Jones' character uses her "real TV voice" to wonder "How old is a grown-up?" or to greet a playmate "I like to meet new friends. Let's play with Mrs. Beasley!"

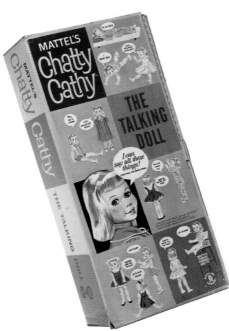

CHATTY CATHY
Mattel 1960

Chatty Cathy opened the whole show. Her success as Mattel's first talking doll set the stage for the arrival of a multitude of other jabbering jovials. Cathy herself was an original creation unassociated with any pre-existing cartoon or comic character. She befriended a decade full of girls with a sweet sincere voice that beckoned for a cookie, to have her hair brushed, to hear a story, or just to accompany her playmate asking, "Please take me with you."

Chatty Cathy was fully jointed, available as either a blonde or brunette with blue or brown "sleeping" eyes. She originally came packaged with two dresses and an eight page pamphlet-like story book. Later her vocabulary was expanded to 18 phrases, added pig

tails and an auburn hair model, while offerering accessory sets of pajamas with slippers, a party coat and hat, a shorts and jacket outfit, or a doll size bed. As popularity grew, so did the Chatty family with Charmin' Chatty, Chatty Baby, Tiny Chatty Brother, and a singing Chatty. Finally, fate was to change in 1970 starting with a re-designed face. The doll was also shortened by a couple of inches and now only spoke nine phrases. Even though competitively priced, this new Cathy soon lost favor with the public and was discontinued. Despite an unfortunate end, the original has gone into toy history as the big sister of all talking dolls and a generation's favorite freckle faced friend.

CHARMIN' CHATTY
Mattel 1963

This lanky little girl stretching up to two feet tall was one-of-a-kind with her granny glasses and happy-yet-homely expression. She was posable, had "sleeping eyes," and came with either blonde or auburn hair. Taking a twist to the pullstring scheme of things, Charmin' had a series of five records to be inserted for 120 varied responses. Promoted as the doll who plays with you, a number of "Let's Play" clothing and accessory sets were available with records based on themes like a birthday party, going shopping, playing nurse, or having a pajama party. A "Travels 'Round The World" gift play set came with foreign language records so Charmin' could chat in English, French, German, Spanish, Italian, Russian, and Japanese.

Expanding play, a pair of specially designed 2 in 1 Chatty games were produced with a Chatty record for responses to the game. Finally, if you want to do the talking for a change, sit Charmin' down and read her own Little Golden Book story.

One other claim to fame that would make even Barbie jealous was Charmin' being chosen as "Cover Girl" for the 1963 issue of *Post* magazine highlighting the holiday season's new toys. Now before I close, let's pull the string once more to see what other future aspirations she may have ... "I'm the President of the USA!"

GIDDY UP!

MR. ED
Mattel 1962

"Isn't it silly? ... talking to a horse."
Not for me. I consider this talking hand
puppet the ultimate Mr. Ed item from one of
the decade's most influential pop culture stars
in TV Land. And what a range of emotions!
Nonchalant: "Just call me Mr. Ed." Anger:
"Who has a horse face?" Romantic: "My
girlfriend has a pony tail." Philosophic:
"I'm a horse of course."

It has long been a challenge for shoppers entering
discount department stores to get their kids in the door
past the mini-arcade amusement area without stopping.
This phenomenal playground is an extravaganza of
revolving hot dogs, hot buttered popcorn, soft pretzels,
Slurpees, video games, and at least one or more coin-op
rides. These riders range from the rocking horse to the
rocking boat, the rocking car, rocking helicopter, or
rocking robot. Whatever it is, you feed in quarters and
it rocks.

The home version rocking horse is a perennial
favorite as were anything else kids could climb in or
on for a ride. Here a few spiffy samples for kiddie
cowpoke transportation plus one other nag.

MARVEL THE MUSTANG
Marx 1967

This pinto pony took tykes for a trot
across the gamelands by tugging at the
handle grips and stepping on foot posts.
Marvel would actually walk along at the
expense of kid power with spurs included
for stubborn steeds.

TONY THE PONY
Marx 1963

Tony the riding Pony takes the
equestrian theme for a cruise. Powered by
one of those gigantic 6 volt batteries, Tony
is propelled by a foot pedal; toe down to go
forward and heel back for reverse. Handle
grips turn the horse's head for steering. This
toy and his television commercial jingle are a
well remembered favorite according to our
recent "gallop poll."

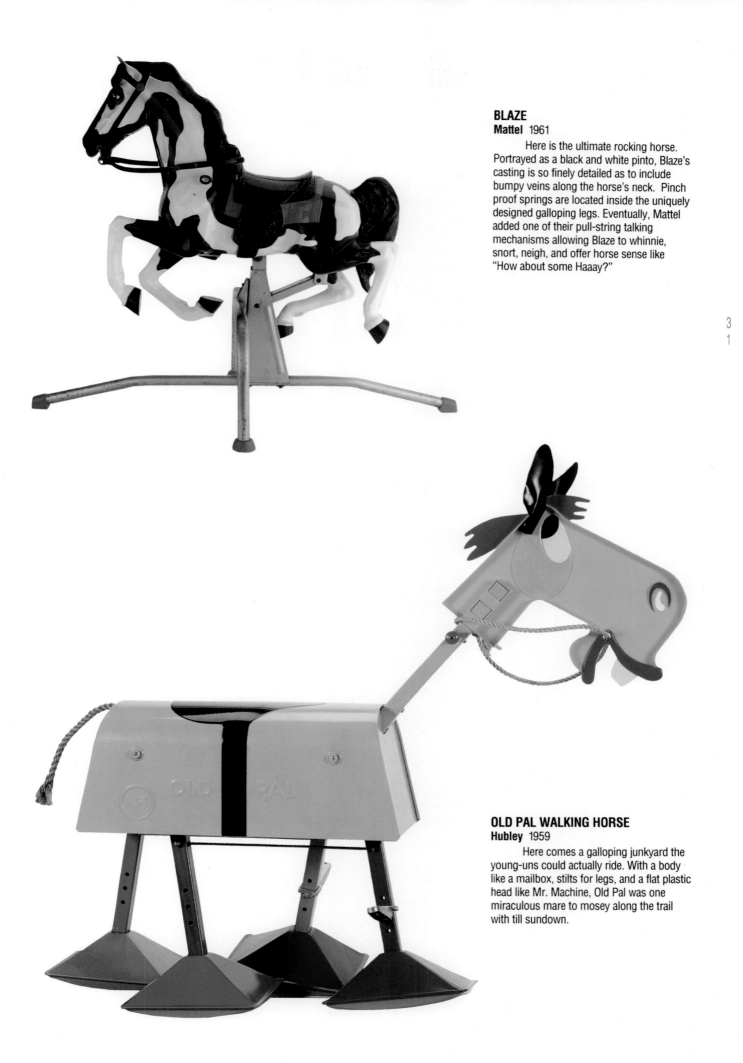

BLAZE
Mattel 1961

Here is the ultimate rocking horse. Portrayed as a black and white pinto, Blaze's casting is so finely detailed as to include bumpy veins along the horse's neck. Pinch proof springs are located inside the uniquely designed galloping legs. Eventually, Mattel added one of their pull-string talking mechanisms allowing Blaze to whinnie, snort, neigh, and offer horse sense like "How about some Haaay?"

OLD PAL WALKING HORSE
Hubley 1959

Here comes a galloping junkyard the young-uns could actually ride. With a body like a mailbox, stilts for legs, and a flat plastic head like Mr. Machine, Old Pal was one miraculous mare to mosey along the trail with till sundown.

Give me your scribblers, your ink spillers, and paste eaters. Bring on those scissors snipping, glue globbing crafters of tomorrow who never made a mistake that couldn't be painted over.

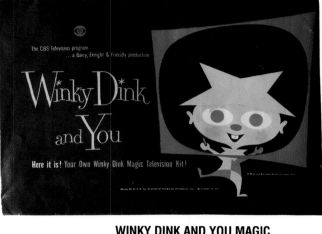

WINKY DINK AND YOU MAGIC TELEVISION KIT
Standard Toykraft 1950's

Here's the official kit which gave kids permission to draw directly on their TV screen. Simply attach the smooth sticky vinyl Magic Screen to your television and follow Winky Dink's coloring instructions. A cloth could easily wipe off waxy Magic Crayons for repeated use. Problems arose with the artistic expressions of scribbling kids, deprived of the Official Magic Screen, drawing directly on their television sets.

WINKY DINK INFLATABLE
Ideal 1950's

True to the TV character's M.O., this inflatable comes with interchangeable *Colorforms*-type face pieces requiring kids to create and re-create Winky Dink's face.

MISS FRANCES DING DONG SCHOOL BELL
The N.N.Hill Brass Co. 1950's

PLAY-DOH
Rainbow Crafts Inc. 1960

Along with their own Pixie mascot, Play-doh licensed Captain Kangaroo to sponsor their yummy-smelling, salty-tasting, squeezy-soft modeling compound. Back then, kids were suggested to mold the colorful clay into a clown, fruit, animals, jewelry, and of all things, an ashtray!

OFFICIAL DING DONG SCHOOL FINGER PAINT
Standard Toykraft Products 1958

Here's Miss Frances giving kids the goods to goosh their fingers through colorful paints and smear them all over papers. The beauty of this is, not only that you don't get yelled at, but your picture gets hung on the refrigerator for all to behold! Sounds too good to be true.

SNIPPY ELECTRIC SCISSORS
Ungar 1950's

This original fish design handle with bulgy eyes allowed kids to safely buzz through paper projects without fear of cutting or poking each other.

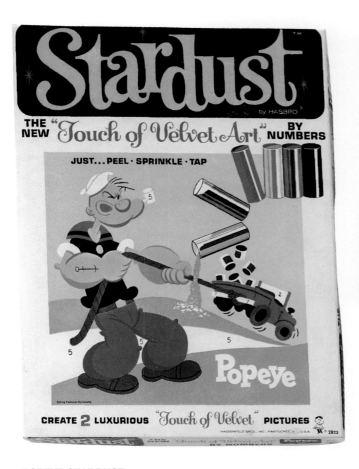

POPEYE STARDUST
Hasbro 1965

This is slick. First peel off numbered sticky sections from your picture. Then pour tubes of corresponding colored flock over the surface. Give it a little tap, shake, feathering and voilá, velvet art! Now see if Mom will let you hang it in the living room.

ETCH-A-SKETCH
Ohio Art 1960

Who would have ever thought such a basic idea as twisting a couple of knobs to form vertical or horizontal lines on a gray screen could be so addictive? So much so that several decades and over 50 million units later, the original configuration is as popular as ever. In fact, as we speak, current production demand continues with the assembly of 8,000 Etch-A-Sketch units per day. What I don't get though, is why my pictures never look like the ones on the package?

WOOLY WILLY
Smethport Specialty Co. 1950's

Here he is, the man with the magnetic personality who made hair replacement happy. At the wave of a wand, we either giveth or taketh away. The popularity of Willy's timeless transplants has touched five decades so far. And to think he's never even had an info-mercial!

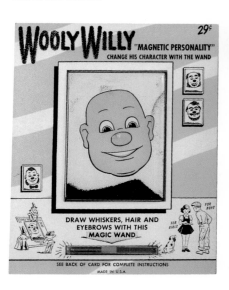

PRESTO SPARKLE PAINTING SET
Kenner 1962

Kenner's popular Presto Sparkle sets featured a number of characters including this one with *Rocky and Bullwinkle*. First, paint the pictures using solid water colors in plastic tubes without water or brushes. Then add sparkle to your work brushing on pre-mixed jewel-like gluey glitter. Even though this was basically a no-muss no-fuss set-up, sparkle usually found its way to fingers and foreheads. So while kids took a shine to their finished artwork, it put a shine on them.

READY MISTER MUSIC?

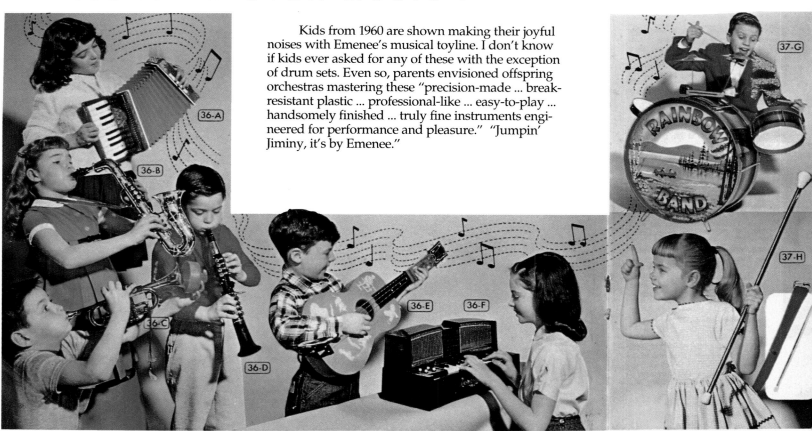

Kids from 1960 are shown making their joyful noises with Emenee's musical toyline. I don't know if kids ever asked for any of these with the exception of drum sets. Even so, parents envisioned offspring orchestras mastering these "precision-made ... break-resistant plastic ... professional-like ... easy-to-play ... handsomely finished ... truly fine instruments engineered for performance and pleasure." "Jumpin' Jiminy, it's by Emenee."

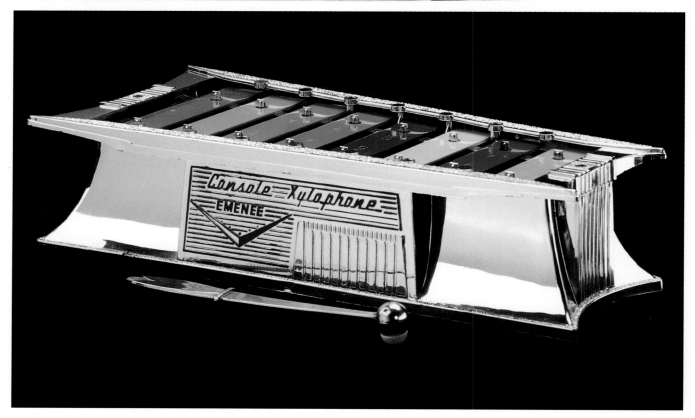

CONSOLE XYLOPHONE
Emenee 1959

Sure everyone Do-Re-Mi'd on kiddie xylophones, but lucky were those who tapped out their color coded tunes on Emenee's cool chromey console.

ELECTRIC GOLDEN PIPE ORGAN
Emenee 1958

A twist of the switch, the whir of an electric motor, and the touch of plastic keys put Golden Pipe organ notes into the air. So that those notes could pass for music, a paper strip was placed along the keyboard identifying the keys for mini-maestros without musical training. Now the entire household could be subjected to er' I mean treated to timeless favorites like *Mary Had A Little Lamb, On Top Of Old Smoky,* or *Twinkle Twinkle Little Star.*

HOME ON THE RANGE

IT ALL STARTED WITH A POTATO

Since 1952, Hasbro's Mr. Potato Head and his smorgasbord entourage of funny face fruit and vegetable friends have been teaching kids to play with their food. Now after four decades and over 50 million Potato Heads sold, it seems our perennial potato remains headstrong about his hilarious heritage. Therefore we thought it appropriate to butter him up with a look into his early days as a young spud.

TOOTY FROOTY FRIENDS
Hasbro 1964

This series was the first to come with their own soft plastic heads thus eliminating the need to invade the grocery order. They were sold either individually with a Mr. Potato Head or as a group called the Tooty Frooty Friends. Our roster includes Pete the Pepper, Katie Carrot, Cooky Cucumber, and Oscar Orange.

PICNIC PALS
Hasbro 1966

Following in the Frooty's footsteps, the Picnic Pals also came with host, Mr. Potato Head but also included a third drink or condiment character. The name Picnic Pals was reserved for the group set featuring: Frenchy Fry with Mr. Soda Pop Head, Frankie Frank with Mr. Mustard Head, and Willy Burger with Mr. Ketchup Head.

DUNKIE DONUT HEAD
Hasbro 1960's

Dunkie is a *Styrofoam* Donut Head friend of Mr. Potato Head exclusively licensed by *Dunkin Donuts* as an advertising premium.

MR. AND MRS. POTATO HEAD CAR AND TRAILER
Hasbro 1950's

These early *Styrofoam* head sets recommended use of real fruit and veggies.

MR. AND MRS. JUMPIN' POTATO HEAD
Hasbro 1966

In addition to face changing festivities, this curious couple operate as wind-ups shaking and wiggling their necessary accessories.

YUM YUM EAT 'EM UP

Throughout Childhood's Menu of Play, did you or any of your friends ever turn down the opportunity to play with a food toy? Second question: Was there ever a food related toy with ingredients remaining the day after you got it?

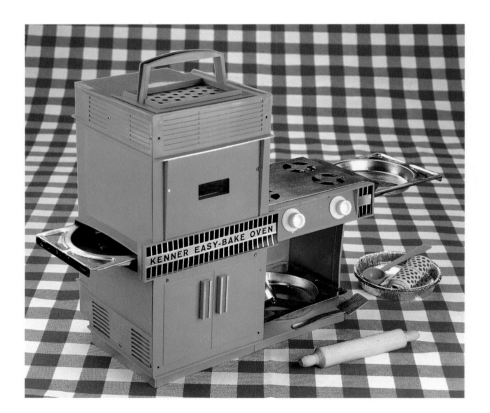

EASY BAKE OVEN
Kenner 1964

"Bake your cake and eat it too!" This boxy bulky light bulb oven was and still is sacred to the former 60's girls who once mixed up Easy Bake recipes "Like Mom" including brownies and snow mounds, apple pie, pizza, chocolate fudge, pretzels, and crazy cake. Accessories include three slide-through baking pans, measuring spoon, rolling pin, mixes and more!

FROSTY THE SNOWMAN SNO-CONE MACHINE
Hasbro 1961

Crushed ice and Dixie cups! Grape and cherry syrup! Frosty and his famous snow cones spanned the decade in kitchens across the nation. Over the years a few changes were made in packaging, syrup bottle colors, and inclusions such as tray molds and candy sprinkles. A deluxe set later included a vendor's hat, apron, and poster for your Sno-cone stand.

Frosty Sno-Man®
SNO-CONE
MACHINE
by Hasbro®
10 different delicious
FLAVORS
TASTY TREAT
SNO-CONE

SQUIRT
The Squirt Co. 1961

ELECTRIC FOOD CENTER
Ideal 1961

In the 50's and 60's, the word *electric* was used as a buzz word to promote toys. Kids normally shunned from electricity were now permitted to unleash its power with a toy that apparently does something *automatic*. In this case, a central motorized deck could interchange a blender, mixer, or ice cream machine. A timer in front could assure proper stirring for recipes like Super Shake, Cherry Flip, Chocolate Egg Cream Soda, or Patti Playpal Punch.

MARKY MAYPO
1950's

Marky peeks in the mixer bowl hoping to find some of his favorite cereal.

MR. MIX-IT
Ideal 1961

Pumping Mr. Mix-It's derby up and down mixes your shake with "whirling action." Then insert a straw and sip. Hopefully if you didn't over-stir, there's still some chocolate syrup waiting at the bottom of the glass. Yummmmmm.

CANDY CIGARETTE BANK
Miner-Matic 1960

Don't look for an updated version of this in today's toy stores. Before the health hazards were fully realized, even the Flintstones did cigarette commercials. Therefore why not pretend to puff sugary sweet candy cigarettes just like Mom, Dad and the cowboy on TV.

COCA COLA TOY SODA FOUNTAIN
Trim Molded Productions 1953

"Hey Kids! It really works!" Insert a bottle of Coca-Cola inside the soda fountain and connect the handle/spout. Now you're ready to dispense directly into four miniature Coke glasses. Not just a toy, "It's the real thing!"

GUM MACHINE AND BANK
Hasbro 1961

The technique not listed on the directions is to leave the back open and use the same penny over and over to buy out the machine's entire supply. Odds are good if you follow the trail of empty *Chiclets* boxes they'll lead to kiddie culprits with jaws full of gum.

SPEEDY ALKA-SELTZER
Miles Laboratories 1950's

INCREDIBLE EDIBLES
Mattel 1966

Plug in your Sooper Gooper and pour in the Gobble-Degoop. Cook up a batch of cinnamon flowers or root beer turtles, a butterscotch gingerbread man or licorice octopus. You can make everything from lizards to butterflies in six sugarless incredible flavors, all completely edible.

FARFEL HAND PUPPET
Juro 1950's
Jimmy Nelson's famous jaw flapping Farfel checks out the competition to N-E-S-T-L-E-S whom he believes makes the very best Choc ... late.

COCOA MARSH SODA FOUNTAIN
Super Syrups Inc. 1958
"Here it is, a real soda fountain!" If your 'fridge is well stocked, this toy really has working pump bottles and compartments to contain syrups, cherries, nuts, sprinkles, and ice cream. A 12 oz. jar of chocolate *Cocoa Marsh* comes "free inside" to make recipes like the Cola Coola , Cake Burger, and Gee-Whiz Fizz.

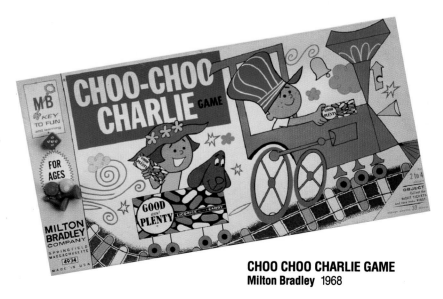

CHOO CHOO CHARLIE GAME
Milton Bradley 1968

FUNNY FACE DRINKS
The Pillsbury Co. 1969
These mugs were made as "send away" premiums for the famous powdered mix soft drinks with wacky names and faces to match. Their ridiculous roster reads Freckle Face Strawberry, Lefty Lemon, Choo Choo Cherry, With-it Watermelon, Goofy Grape, Jolly Olly Orange, Frooty Tooty Fruity, Chug-A-Lug Chocolate, and Loudmouth Punch.

PLAYING HOUSE

Unlike so many other areas of Toyland that miniaturize life into pretend playthings, the themes of cooking and cleaning were often re-enacted with real working replicas like sweepers that sweep, blenders that blend, washers that wash, ovens that bake real pastries ... make a real mess ... and were real fun!

CANDY FASHION
Deluxe Reading 1961

Here's Candy Fashion for sweet little girls who won't need to make multiple toy store trips to expand their doll's wardrobe. Why? Because all in one big glorious box, comes Candy Fashion with four coordinated outfits, three dress forms, and other accessories like purses, shoes, hats, and my favorite ... cat's eye sunglasses!

SUZY HOMEMAKER STEAM IRON
Topper 1966

Just add water, plug in, and Sssssssss ... one *impressive* toy.

DREAM KITCHEN
Deluxe Reading 1963

This 176 piece "Dream Kitchen" is loaded with special effects, accessories, and what looks like ... good eats! Pop in a battery and a rotisserie revolves your turkey. Meanwhile on the range top, pretend burners light to a glowing red. Even running water is provided for the sink and dishwasher from hidden overhead tanks which simply disperse downward. Now swing out some chromey refrigerator shelves to reveal a fun feast of food and drink while other packaged playfood gets stored away in cupboards. A table and chairs matching the appliance colors complete the set with dishes and silverware. Tell your dolls to come hungry.

BETSY WETSY
Ideal 1956

Like responding to a word association test, anyone from this era who hears the name, Betsy, automatically thinks ... Wetsy. Available as either a molded hair infant or a larger rooted hair toddler, Betsy's claim to fame is to make the water in her bottle disappear and then re-appear in her diapers or as tears. She also has sleeping doll eyes and coos when squeezed. Besides her clothing, other accessories include real soap, baby powder in a tin container, wooden clothes pins, and a glass drinking bottle.

DOLL HOUSE

Playing house with this lovely decorated seven room Doll House will be heaps of fun. It has 27 colorful pieces of furniture and even a swiming pool and a water slide. By T. Cohn Inc. Ask for 21BR1

T. Cohn Doll House ad from Billy and Ruth, America's Famous Toy Children 1951

SPEED PHONE
The Gong Bell Mfg.Co. 1950's

SUBURBAN COLONIAL DOLL HOUSE
Marx 1950's

I think there was a law that every girl from the 1950's and 60's must own a tin lithographed doll house in order to pass grade school. An endless variety of these generic metal mansions were equipped with a zillion household accessories like tables and chairs, TV's and sofas, refrigerators, and even the bathroom commode. A unique feature to this *Leave it to Beaver*-style two-story was a nursery with Walt Disney character decor.

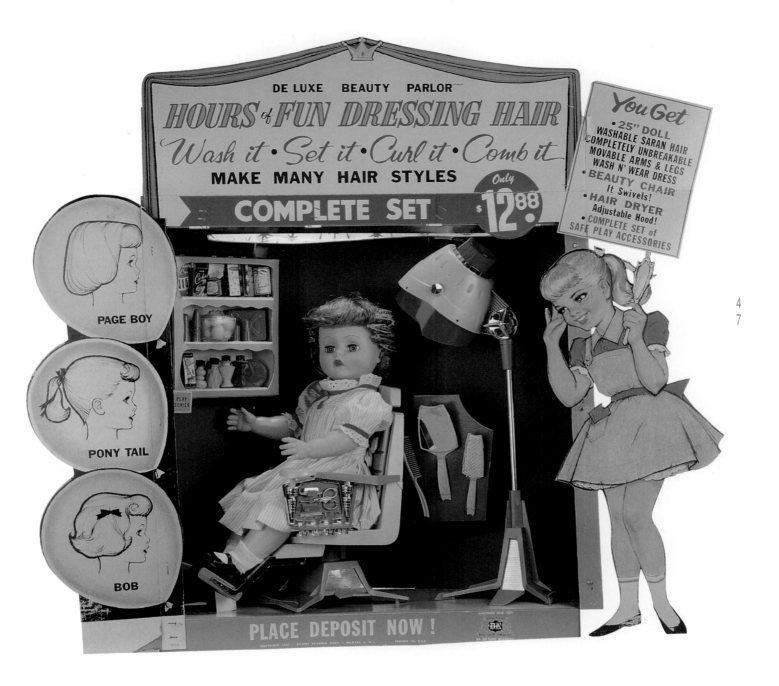

BEAUTY PARLOR
Deluxe Reading 1961

Need a hairdo? The Deluxe Beauty Parlor is equipped to do Dolly's make over. Besides featuring the swiveling beauty chair and adjustable hair dryer, other accessories include a hand mirror, brush and comb set, little hair curlers, cotton puffs, pretend lipstick in Rosebud or Dreamy Pink, perfumes and powder, plus everything you need for a miniature manicure. Suggested hairstyles include a Page Boy, Bob, or Pony Tail ... What, no Beehive?

LITTLE AUDREY DOLL
Gund 1960's

LITTLE AUDREY BREAKFAST TIME
Brothers Mfg.Co. 1962

Not only did Little Audrey endorse this set of kitchenware but even the play food officially pretended to be *Kelloggs Corn Flakes*, *Rice Krispies*, and *Domino Sugar*.   *Little Audrey Breakfast Time and Little Audrey doll appear courtesy of Ira Gallen*

LITTLE RED SPINNING WHEEL
Remco 1961

Grab a hank of wool, some needles and spools, and this near two foot high "little" spinning wheel is supposed to make an infinite variety of household necessities. Some suggested ideas include pot holders and pocket books, hook rugs and hats, or even clothes for dolls. Remco claimed "for girls it's great!" and yet, believe it or not, they also made a smaller version of this toy called Davy Crockett's Spinning Wheel. You heard right. So may I part with one last bit of advice for all the "complete-ist" Crockett collectors ... I think it goes, "Knit one then pearl two."

SCOOBA DOO
Mattel 1964

Our Mod fashioned scat-singing Scooba Doo plays it cool with the pull of a string. "Hey Doll, like you're way out," or "Come on let's get with it, like Wheeee!" Her favorite two pastimes seem to be music, "I dig that crazy beat ... Yeah!" and surprisingly ... "I dig food, like when do we eat?" For anyone who doesn't get her lingo, Scooba Doo offers an explanation "I'm hip, like you know ... a Beatnik!"

SUZY SMART
Deluxe Reading 1961

For those who couldn't get enough of school, Deluxe turned work into play with Suzy Smart. Along with the doll, a toy school desk and blackboard were supplied with a pencil, chalk, and eraser. Suzy appears to be the twin sister of Deluxe's Beauty Parlor doll with the added feature of recorded speech. At the touch of a chest button Suzy introduces herself before showing some smarts with spelling, arithmetic, and finally reciting *Twinkle Twinkle Little Star*.

If you were going to count all the Barbies ever made on fingers and toes you would need the help of 25 million people. For according to Levy and Weingartner from *Inside Santa's Workshop*, over 500 million Barbies have sold since the doll's introduction in 1959. Then don't stop there. The on-going sales for the world's most popular fashion doll continue to ring or beep cash registers at the rate of one every two seconds. Barbie attained this achievement with a little help from her ever expanding circle of friends including Ken, Midge, Allan, Skipper, Skooter, Christie, Francie, and others who if lined up head to toe would create 3 1/2 more circles ... around the Earth!

BARBIE SPORTSCAR
Irwin 1962

The idea of licensing a Sportscar for Mattel's young lady on the go turned toymakers into car salesmen. According to *Toy and Hobby World* Irwin's Barbie Sportscar did $2.5 million dollars in sales during its first year of production.

KEN HOT ROD
Irwin 1963

Here comes Ken cruising the beach in hopes of impressing the number one "Babe in Toyland," Barbie, with his street rod. I hope he can dazzle her with its features better than department store catalogs that boast " rubber-like tires" and "wheels actually revolve." If I was him, I'd try bragging about all the chrome trim and hope she doesn't notice the construction of the car bodies being made of polyethylene plastic (the same material used for bubble bath bottles). Regardless, these two convertibles have the distinction of being the first official licensed vehicles of the famous fashion teens and therefore are coveted by collectors worldwide. Besides, how many cars can drive down to the shore, be overtaken by a tidal wave, and then ... float?

IT'S ALL IN THE GAME

Spinners and dice, markers and marbles sent us into the Gamelands to capture the jewel, deliver milk, catch a mouse, win a watch, become king, or simply be first to reach FINISH!

GIANT WHEEL COWBOYS AND INDIANS
Remco 1958

The early Giant Wheels were designed with Remco's ornate foil plate trim before switching to paper labels. The play board was also a giant sized rolled mat to race the skinny figure markers to the big finish! Other Giant Wheel theme games include Thrills and Spills Horse Race, Hot Rod Sport Car Race, Mississippi Speed Boat Race, Picture Bingo, and Old Maid.

SPUDSIE THE HOT POTATO
Ohio Art 1967

TIME BOMB
Milton Bradley 1965

KING OF THE HILL
Schaper 1963

STICKY FINGER
Mascon 1965

BASH
Milton Bradley 1965

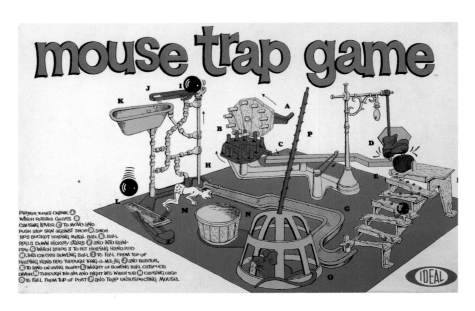

MOUSE TRAP
Ideal 1963

Don't try to re-invent the mouse trap. This one designed by Marvin Glass and Associates has infatuated kids for 3 decades so far. It was just the new twist of a board game Ideal was looking for-or should I say the new kick?-that got the ball rolling, thereby, spring-boarding their sales? The toy was such a hit with kids, Ideal announced having sold 1,200,000 Mouse Trap Games in the first year netting some not-so-cheesy profits.

CRAZY CLOCK
Ideal 1964

What's crazy about this sequel game to Mouse Trap is that it flopped. Re-using Mouse Trap's successful formula of a Rube Goldberg style wacky gizmo, the series of actions eventually spring a sleeping man from bed. So, what went wrong? Everyone kept buying Mousetrap.

FISH BAIT
Ideal 1965

Ideal went fishing with one last attempt to duplicate the success of Mouse Trap. This time, however, it's the fish who gets to tell the story in the end. Once the trap is ready and someone's fishermen is sitting on the dock of the bay, they're hooked! The contraption is set-off and, providing everything works, they get pulled into the giant mouth becoming "Fish Bait!"

CONEY ISLAND PENNY MACHINE
Remco 1959

Like in a Penny Arcade, you gaze at the revolving turntable of temptuous trinkets. Peppered amongst junky rings, marbles, whistles, and hollowed out vehicles were always a few of the more desirable delights: a compass, watch, or harmonica. While nervously waiting your turn, the big decision of which to try for is tormenting ... until, You're next! Insert the coin and grab the levers. This is it!

Too intense? Don't worry. The big advantage to Remco's Penny Machine is the ability to continue playing as long as it takes to win the best bauble.

TUMBLE BUG
Schaper 1959

Watch these little buggers flip flop down hill switching lanes or bumping each other ... whatever it takes to win.

COOTIE
Schaper 1949

A 1956 mini-comic brochure called *The Friendly Cootie Bug* promotes Schaper's toyline while telling the story of the saddest little bug in Insectland. A good fairy finally grants his wish to become a new and special toy Cootie Bug which, according to the ad, "belongs in your home."

SKUNK
Schaper 1953

In this dice game that promises "thrills, suspense, excitement, and laughter" you can get points and chips, or you can get skunked.

TICKLE BEE
Schaper 1958

"It wiggles and jiggles from flower to hive, this magnetic Bee seems almost alive."

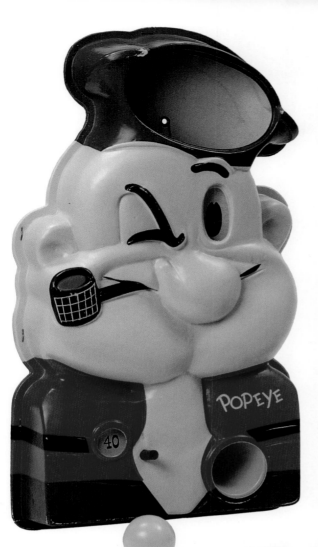

POPEYE CHECKERS
Ideal 1959

POPEYE BALL TOSS
1968

**YOGI BEAR SCORE-A-MATIC
BALL TOSS**
Transogram 1960

PALADIN CHECKERS
Ideal 1960

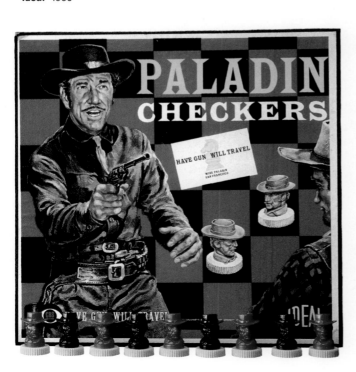

VIDEO VILLAGE
Milton Bradley 1960

On television, the contestants actually became the markers walking along a life-size game board. Before reaching "Finish," they would cross the bridge to win cameras and coffee pots, watches and waffle irons, roller skates and skis, go to jail (bummer), or get sacks of money! One big plus to the "home version" was not having to pay "Prize Tax" on your winnings.

THINK-A-TRON
Hasbro 1961

Put aside your Think and Do book, it's time to Think-A-Tron. Turn the crank. There's buzzing and lights flashing as the electronic question and answer computer "thinks." Finally, a lighted response indicates an A,B,C, T, or F to one of 300 questions selected from the Book of Knowledge Children's Encyclopedia.

HASBRO

SEE ME ON TV

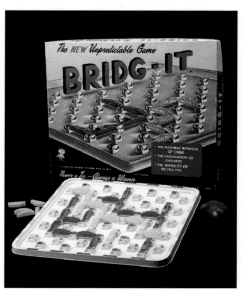

BRIDG-IT
Hasbro 1960

This contest of strategy and a little luck is basically a race across the board, building while blocking your opponent with little plastic bridges. "Never a tie. Always a winner!"

MYSTERY DATE
Milton Bradley 1965

"He's here! My Mystery Date!" This game promising "romance and mystery," involves opening a mystery door to reveal a date who could either be a Dream or a Dud. So is it against the rules to slam the door on the Dud? Better not. I heard he's now making six figures and has a summer home in Martha's Vineyard.

MISS POPULARITY
Transogram 1961

This true American Teen game selects a girl with Popularity Bulletin Boards, an Automatic Answer Phone, and Judges to be Miss Popularity! The winner is awarded money and a loving cup but that's all. If you want a "Dream Date" that is yet another board game.

THE OFFICIAL BEATLES
Remco 1964

    From tots to teens, everyone's favorite Fab Four were John, Paul, George, and Ringo, who lead a "British Invasion" in a Yellow Submarine. "Koo Koo Ka-Choo."
(See Epilogue)

THE BEATLES FLIP YOUR WIG GAME
Milton Bradley 1964

THE MONKEES GAME
Milton Bradley 1967

THE MONKEES TALKING HAND PUPPET
Mattel 1966

    Hey Hey it's the Monkees! It seems TV's groovy foursome aren't too busy to put each other down with a lot of laughing, jeers, crashing, and wild animal noise impersonations. "Come on guys, quit monkeying around ... Aw, shut up."

CAPTAIN KANGAROO
Baby Barry 1950's

    Since the 50's, armies of children were befriended by Bob Keeshan, who portrayed a very kind Captain of Kid-dom. Inside his headquarters, the Treasure House, daily misadventures took place with Mr. Green Jeans, Grandfather Clock, Dancing Bear, Bunny Rabbit, and Mr. Moose.

CAPTAIN KANGAROO GAME
Milton Bradley 1956

THE SUPERMAN GAME
Merry Mfg. 1960's

THE ADVENTURES OF POPEYE GAME
Transogram 1957

MY FAVORITE MARTIAN GAME
Transogram 1963

THE BATMAN AND ROBIN GAME
Hasbro 1965

I DREAM OF JEANNIE
Libby 1966

Girls loved the idea of folding arms and blinking their dreams true. Guys never tired of hearing Jeannie's constant quote "Yes, Master!"

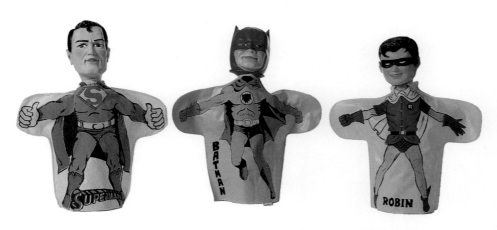

SUPERMAN HAND PUPPET
Ideal 1965

BATMAN AND ROBIN HAND PUPPETS
Ideal 1966

I DREAM OF JEANNIE GAME
Milton Bradley 1965

POPEYE TALKING HAND PUPPET
Mattel 1967

"Spinach don't fail me now." Too bad for our bulgy armed buddy the can of spinach sticker pre-dates "scratch and sniff."

MERRY MILKMAN
Hasbro 1950's

Be first to load your dairy truck with little wee plastic milk bottles, eggs, and butter, then deliver them to your neighborhood. Beware the spinner position, Cracked Bottle, Return to Dairy when playing this "Exciting Game and Toy."

FEARLESS FIREMAN
Hasbro 1950's

I always thought firemen were in a hurry until trying to play this sequel to Merry Milkman. Before rescuing little plastic people from a burning building, a series of spins are required. First, spin for a Fire Alarm, again for a Fire Engine, next a Rush to Building, followed by either a Ladder to Window and Carry to Safety, or Jump to Safety using the net. Beware the Overcome by Smoke spinner position while playing this "Thrilling Game and Toy."

POLICE PATROL
Hasbro 1950's

Hasbro takes us downtown to play this third and final " Action Game and Toy." This one has you selecting from 3 spinners and a deck of cards to direct traffic, give citations, help lost children, chase robbers, call headquarters, or cruise in a patrol car all while scoring points. Or instead, the directions suggest using it as a toy, "enjoy many hours of creative play combining Merry Milkman, Fearless Fireman, and Police Patrol to make a complete community. Your child will find extra pleasure from playing with all three."

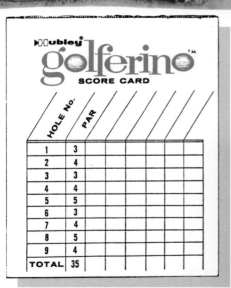

GOLFERINO
Hubley 1962

As you can see, Mr. Golferino knows the first and foremost rule to golf: wear loud clothing and a funny hat. Besides his flair for Fairway fashions, he hunkers down and keeps the old eye on the ball. The rest is up to you. Begin by turning a knob to position the pudgy putter. Line up your shot and push a lever. Fore! With just the right swing, the ball rolls through various obstacles or hazards, lands on the green, and sinks in for a hole-in-one. Your final stroke finishes the course ringing a bell and pops up a loving cup trophy.

Hubley golferino SCORE CARD

HOLE NO.	PAR					
1	3					
2	4					
3	3					
4	4					
5	5					
6	3					
7	4					
8	5					
9	4					
TOTAL	35					

JUNGLE HUNT
Hubley 1963

Are you ready to go on a "Big Game" shootin' safari with Hubley's Jungle Hunt? First, set the 60 second timer and hit the switch. A bell will sound when the time is up. Now watch as moving animals randomly pop in and out of scenery. The hunter takes aim controlled by a swiveling pistol grip which also raises his rifle, and clicks a shot. If the angle and timing are correct, our ferocious friends fall over. Quick here comes one now! ... Ding!

SAY KIDS! WHAT TIME IS IT?

DOIN' THE HOWDY DOODY
Unique Art 1950's

Doin' the Howdy Doody appears courtesy of Rick Rubis

"It's Howdy Doody Time!" Buffalo Bob Smith pioneered children's programming and television itself with a show format mixing live actors with a cast of marionettes before bleachers of boisterous boys and girls known as the Peanut Gallery. With some 15 million viewers at home, demand for those seats filled a waiting list so long that, according to Shulman and Youman's *How Sweet It Was*, expectant mothers were known to have signed up unborn future fans. Naturally, this level of popularity carried "over the counter" the minute Howdy Doody character merchandise arrived. A 1955 Kagran Corp. catalog displays a diverse deluge of Doodyville while proclaiming "It's a Howdy Doody World! ... in which youngsters can play with Howdy Doody Toys, listen to Howdy Doody records, read Howdy Doody books, wear Howdy Doody apparel," and at meal-times "have Howdy Doody food products adorn their table." This is all done in accordance with the show's "wholesome creed ... of fun and fantasy, devoid of any elements of violence." Furthermore, Kagran mentions Howdy Doody even being approved by churches and the PTA. Now for those who are befuddled as to how Buffalo Bob had so much luck with everyone's favorite freckle-faced character, consider this: All he had to do is knock on wood!

HOWDY DOODY TV GAME
Milton Bradley 1955

HOWDY DOODY MARIONETTES
Peter Puppet Playthings 1950's

Pint-sized puppeteers flipped over Buffalo Bob's famous characters including Howdy Doody, Clarabell, Dilly Dally, Princess Summer Fall Winter Spring, Phineas T. Bluster, and flying in from above, Flub-A-Dub.

Howdy Doody TV Game appears courtesy of Ira H. Gallen

Flub-a-Dub, Clarabell, Dilly Dally, Princess Summer Fall Winter Spring, and Phineas T. Bluster appear courtesy of Bob and Barb Pierce

WHEN YOU WISH UPON A MOUSE

Walt Disney often put his animated empire into prospective with the expression, "It all started with a mouse." Since then, Mickey, Minnie, and friends continue to lead never ending parades of characters into our hearts with their unique style of family entertainment.

More often than not, those characters jumped from animated cells into a toy box which earned Walt a seat in the Toy Hall of Fame. Here is a sampling of 50's and 60's items from the *Mickey Mouse Club* to *Walt Disney's Wonderful World of Color*.

WALT DISNEY'S BABES IN TOYLAND
Colorforms 1961

WALT DISNEY'S MOUSE GE-TAR JR.
Mattel 1955

WALT DISNEY'S MICKEY AND MINNIE MOUSE HAND PUPPETS
Gund 1960's

WALT DISNEY'S MOUSKETEER HAT
1960's

WALT DISNEY'S ZORRO, A FUN GAME FOR ALL THE FAMILY
Whitman 1958

WALT DISNEY'S SLEEPING BEAUTY HAND PUPPETS
Gund 1960's
    Merryweather, Flora, Fauna, King Stefan, King Hubert, Prince Phillip, Princess Sleeping Beauty, and Maleficent.

WALT DISNEY'S ANNETTE CUT-OUT DOLL
Whitman 1960

WALT DISNEY'S MOUSKETEER SOAKY
Imco 1960's

WALT DISNEY'S OFFICIAL DAVY CROCKETT INDIAN FIGHTER HAT
Weathermac 1950's

WALT DISNEY'S SNOW WHITE SOAKY
Imco 1960's

WALT DISNEY'S PETER PAN
Sun Rubber Co. 1953

WALT DISNEY'S SWAMP FOX
Parker Brothers 1960

KING OF THE WILD FRONTEIR

In 1956 a young actor named Fess Parker was hired by Walt Disney Productions to portray the legendary American hero, Davy Crockett. Wearing buckskins, a coonskin cap, and balancing an ornate Flintlock in his forearms stood the man who was to personify the soft spoken backwoods hero who "killed him a bear when he was only three." And if anyone on Earth wasn't already acquainted with the deeds that made Crockett, "King of the Wild Frontier," that would soon change. Immediately following Disney's television programs, a new King was crowned, "Fess Parker, King of the Davy Crocketts."

From that point in time through the early 60's, all the coonskin caps and Old Betsey rifles from a mountain top in Tennessee along with every other conceivable Crockett product sold over 100 million dollars worth of merchandise. Whether licensed through Walt Disney's Fess Parker character or generically based on the historical figure, everyone flipped over this favorite Frontiersman.

DANIEL BOONE
Remco 1964
Even though this buckskinned figure of Fess Parker was packaged as Daniel Boone, the exact same toy a decade earlier would have been a hot selling Davy Crockett toy. So my advice to Crockett fans ... hide the box.

DAVY CROCKETT
Breyer 1950's

THE ALAMO PLAY SET
Marx 1960
Here's one toy history won't let you forget ... "Remember the Alamo!" When I hear that battle cry my mind replays the legendary image of Davy Crockett swinging his flintlock through the smoky mist of his last battle.

In 1955, Louis Marx re-created another unshakable image, the Alamo play set. Initially, the toy was licensed as Walt Disney's Official Davy Crockett at the Alamo. The early set is distinguished by a tin lithographed gate and an official Davy Crockett character figure. Then in 1960, inspired by John Wayne's epic film *The Alamo*, the play set was revamped without licensing. The front gate was redesigned in plastic and the official character figure removed. Undaunted, kids simply substituted one of the frontiersman figures posed clubbing with a flintlock to rewrite history; knocking down any and all attackers of the beloved shrine.

Zorro, the bold renegade who carved Z's with his blade, was portrayed once and for all by *Walt Disney Studios'* Guy Williams. As the mysterious hero, Williams donned black cape and mask to gallop through the night as lightning strikes a Z into the stormy sky. That lightning was actually striking twice for Disney reprising the rage experienced by Davy Crockett several years prior. Soon the ebony equestrian began stacking equally enormous piles of playthings into neighborhood quartels. Only now, all the hats, masks, swords, guns, and games were black and stamped with Z's. Today that popularity rides again as full-grown friends of Zorro who still hear sword swishing Z's in their dreams search for the valuable vigilante into the night ... "when the Moon's shining bright."

WALT DISNEY'S ZORRO SWORD
Walt Disney Productions 1950's

WALT DISNEY'S ZORRO HAND PUPPET
Gund 1950's

WALT DISNEY'S ZORRO
PEZ 1950's

WALT DISNEY'S ZORRO VIEW MASTER STEREO PICTURES
Sawyers 1958

OFFICIAL WALT DISNEY'S ZORRO RIDER AND HORSE
Marx 1958

WALT DISNEY'S ZORRO PUZZLE
Whitman 1957

OFFICIAL WALT DISNEY'S ZORRO PLAY SET
Marx 1958

Setting up a typical scene in one of the Walt Disney television episodes would involve Zorro playing a prank on El Commandant such as replacing the Quartel's flag with a Zorro flag. Then before leaving, the garrison is somehow alerted to his presence and frantically ordered to pursue and capture "the Fox." Marx's 80 piece playset came with 32 Mexican soldiers plus cream colored character figures including Zorro's true identity, Don Diego, his mute friend and accomplice Bernardo, his father Don Alejandro, El Commandant, and "well wisher," Sergeant Garcia.

A day in the life of comedic cave dwellers proved to be a successful formula for over 300,000,000 people in 80 different countries who've met *The Flintstones*. In 1960 Hanna Barbera's sit-com cartoon became a place right out of history as the first animated television series aired during prime time. With complaining creature-powered appliances to Fred's two-foot-drive car, the Bedrock gang entertained the entire family with their sticks and stones community. Instantly, kids and parents alike were Yabba Dabba Dooing along with the Flintstones whose capers of calamity were timeless reflections of our own behavior. During the show's original broadcast years, piles of prehistoric character toys and related merchandise began to erupt from various manufacturers. It is this early batch of Flintstonia that is most sought after by the show's "close friends, bosom buddies, and lifelong pals."

PEBBLES FLINTSTONE
Ideal 1963

Fred and Wilma's daughter came home to play with friends and fans of the show as either a 12 or 15 inch posable doll wearing her cave kid costume and a bone in her rooted hair. The "Tiny" version had related accessories such as a rocking log cradle and a Dino Stroller. Another Baby Pebbles with soft body came tucked in a blanket for beddy bye.

BAMM BAMM AS SEEN ON THE FLINTSTONES
Ideal 1963

Barney and Betty's super strong son was also available as a 12 or 17 inch posable, wielding his famous club ... the better to Bam with.

FRED FLINTSTONE ON DINO
Marx 1962

    If this is really Dino, I guess he grew up. Then in the tradition of his master's trade he apparently has taken up a career at the rock quarry. Starting with bump and go action, Dino moves his tail, legs, neck, head, mouth, and even whistles! All this movement pulls Fred back and forth while taking the ride of his life.

GIVE-A-SHOW PROJECTOR
Kenner 1963

    Bedroom walls and ceilings became movie screens for wiggled productions of Kenner's battery operated slide projector. This year's box premiering Pebbles Flintstone was one of 16 shows including the Jetson's, Huckleberry Hound, Yogi Bear, Quick Draw McGraw, and " all their friends!"

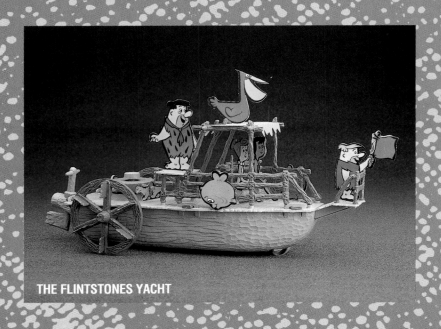

THE FLINTSTONES YACHT

THE FLINTSTONES CAR AND TRAILER

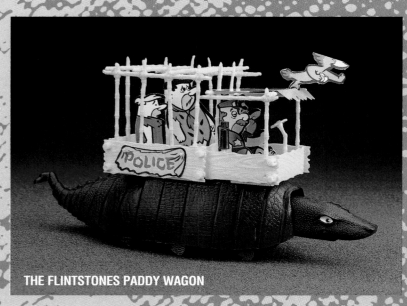

THE FLINTSTONES PADDY WAGON

FLINTSTONES MOTORIZED
MODEL KITS
Remco 1961

In the avalanche of licensed Flintstones' products following the show's first season came this series of prehistoric playthings. The car and trailer, yacht, and paddy wagon were all molded in natural colors making paint unnecessary but required globbing some glue before your battery-ops could head for Bedrock. Punch-out paper people and prehistoric pets quickly slip in to add character.

THE FLINTSTONES PLAY SET
Marx 1961

In miniaturizng the little town of Bedrock, close attention to detail was paid re-creating the characters, their stoney homes, and several other of the town's establishments featuring a super market, gas station, and lunch stand. Other dinosaurs, palms and ferns, and fellow Bedrockers compliment the place right out of history.

COME ON KIDS WIND UP YOUR LIDS

Hot Dog! Some Fun! Oh Joy! In 1962 *The Beany and Cecil Show* splashed on the television screen for "a whole half hour of Bob Clampett Cartooooooooons!" This was a return to the tube for Clampett whose award winning *Time For Beany* puppet show was performed live in the early 50's. However this time, animation proved to be the perfect playground for Bob Clampett's limitless imagination. The cartoon's cast soon grew to over 50 characters including Beeping Tom, Jack the Knife, Slopalong Catskill, and Go-Man Van Gogh.

In the early 60's, a tidal wave of Beany and Cecil toys and games were licensed by Mattel and washed up on stores. Some notables include Beany Copters, talking dolls and hand puppets of Beany, Cecil, Dishonest John, and a Jumping DJ card game which featured many of the show's colorful characters. Following in Mattel's wake, shiploads of other company's Beany and Cecil products ranged from comic, story, and coloring books to puzzles, Leakin' Lena boats, a lunch box, record player, games, and various bathtub items. Stack all these up and it just may reach Beanyland on the Moon!

BEANY COPTER
Mattel 1961
      As the theme song suggests, kids could flip their lids "higher than the Moon!" ... or thereabouts. Another fun feature was to send top secret messages inside the saucer's hidden compartment.

LEAKIN' LENA
Irwin 1962

TALKING DISHONEST JOHN HAND PUPPET
Mattel 1962

If you ever heard a puppet say "Un-hand me you Cad!" it would undoubtedly be coming from "NYA AH AHHHH," Dishonest John. Don't trust his feigned innocence with a line like "You don't think I'm bad do you?" because the next thing you might hear is "AHH, why don't you go soak your head!"

TALKING BEANY BOY
Mattel 1961

True to his character, Beany performs his crowd pleasing classic wimp-out "Help Cecil, Help!" Then other times he's a bit more relaxed and ready to play, "Gee this is fun!" "Let's go exploring," or maybe he'll just ask you to "Tell me a story." If so, remember to keep one eye peeled as Beany warns to "Watch out for DJ!"

TALKING CECIL
Mattel 1961

"Wot the Heck!" It's lovable, gullible, armless, harmless, ten feet tall and wet, Cecil the sea-sick sea serpent to the rescue, "I'm Coming, Beany!" Then in "Just a darn minute" when the coast is clear, Cecil may laugh, slurp a kiss, and say "There goes a good kid."

KOOKIE CAMERA
Ideal 1968

I guess this is the camera to use after playing Mouse Trap. Created by the same inventors, Marvin Glass and Associates, it's not only wacky looking ... it works! Complete with sand clock timer and soup can distorted lens, this actually came with film and took flash pictures. Mod pop-art backdrops made even kookier photos posing behind "fun house" bodies. An exclusive kooky photographer's hood came in the Sears version.

BROTHER AGAINST BROTHER...

In 1861, cannons fired at Fort Sumter in South Carolina in an attack which began America's Civil War. One hundred years later, cannons were again firing , flags unfurled, and muskets cracked fire dividing families and friends. This time "the cause" was for fun and glory as junior cadets replayed the Civil War during its Centennial. The event was commercially commemorated by toys and games, comic books, model and craft kits, costumes, hats, flags, and of course, bubble gum cards. Some items were character authorized from television's *Gray Ghost* and *The Rebel*. Otherwise, the only requirement to interest the public was that it fit the color scheme, blue and gray.

THE JOHNNY REB CANNON
Remco 1961

The television commercial itself was an event to behold. The action begins outdoors with an army of kid confederates in full costume towing Johnny Reb Cannons to the tune of *When Johnny Comes Marching*

Home. The row of cannons are emplaced along a ridge overlooking a life-size union stockade. Then unit by unit they pull the fuse to fire balls at the fort. The battle concludes with a bandaged Yankee and nurse waving.

Remco ballistics report this 2 1/2 foot long field piece to be capable of blasting a single cannon ball, with barrel elevated, a maximum distance of 35 feet. "Harmless" hollow plastic cannon balls were supplied initially but were later replaced with softer dark gray *Styrofoam* balls. The only study not completed regarding the Johnny Reb Cannon is to ascertain who best remembers the toy: those who once fired its ominous ammo or those who got bopped on the noggin.

THE GIANT BLUE AND GRAY PLAY SET
Marx 1961

Our Sergeant's sword raised in the air. His hoarse voice screamed "Fire!" The volley of exploding muskets crackled to a thunderous Boom! The wind was against us blowing the rotten stench of black powder back in our faces. "Reload!" As the smoke began to clear I strained to see if the Rebs were falling back, but instead their battle flag waved even closer. Then their yelling came right out of the fog as they clashed into our forward ranks! ...

This scene took place in Gettysburg at the 1981 battle re-enactment where I first "saw the elephant." Looking back, I wonder how many of the other soldiers began their training years ago with one of Marx's Blue and Gray playsets.

In 1958, a year before the last confederate veteran died, Louis Marx created what was to be their second all-time best selling play set, the Blue and Gray. In 1961, for one year only, a special "Giant" set commemorated the centennial loaded with extra soldiers, a caisson, firing field cannons and mortar, an exploding hill, stone bridge, and hospital wagon. Character figures of Lee, Grant, Lincoln, and Jeff Davis were cast in cream plastic to lead their armies into history. The Giant set's box art cleverly depicts some of the toy's accessories in an illustration based on *Life* magazine's Brandywine Station battle scene. Other special inclusions to this deluxe set were a cardboard battle sounds record, a bullet firing cap gun, and paints. The only feature deleted from the Giant set was a fallen horse and rider offered earlier who has galloped into our photo for posterity.

THE GRAY GHOST GAME
Transogram 1958

Tod Andrews starred as the famed confederate Colonel Mosby leading his raiders across the TV screen and ultimately across the game board into Toyland.

THE REBEL SCATTERGUN
Classy/Kilgore 1960

Johnny Yuma's cap firing double barrel scattergun came with either a black stock or simulated wood color grain for young rebels to settle neighborhood squabbles. A deluxe holster set included the rifle with a pistol, rebel hat, and lucky tooth medallion.

CONFEDERATE SWORD
Made in Japan 1960's

CONFEDERATE PLAY MONEY
Whitman 1960's

GENERAL ROBERT E. LEE, JOHNNY YUMA, AND GENERAL CUSTER
Hartland Plastics 1960's

Hartland Plastics prided themselves on their series of western and historical figures which in most cases bore incredible likeness. The toyline's brochure suggests their use for either play or display. In this trio, Custer and Lee were based on historical photos while television character Johnny Yuma was sculpted as actor Nick Adams.

7
4

Nick Adams starring in The Rebel as Johnny Yuma.

Howdy Pardner! Why don't you set for a spell while I talk about a time when every neighborhood in America was Dodge City. Today, the old western characters and songs seem to exist only in roast beef and steak house food chains. There was a time, though, when western heroes like the Lone Ranger, Annie Oakley, Paladin, and Roy Rogers dominated the television dial. After watching a favorite cowpoke ride off into the sunset, it was time for kids to draw their six shooters, blast a roll of caps, and blow smoke from the barrel before twirling pint-sized pistols into holsters. Now that's something you don't much see any more ... not around these parts. So let's take a gander at some of those toy western wonders, some cowboys and cowgirls, their guns and games, spurs and lassos, and Wagons Ho!

LONE RANGER & SILVER, TONTO & SCOUT
Hartland Plastics 1950's

Since the Lone Ranger was continually having to explain how his mask stood for justice, I'm surprised Tonto never interjected, "Umm Kimosaube, this mask thing not working. Better give it up." Really it did work as the popularity of Clayton Moore and Jay Silverheels' heroic portrayals kept everyone humming *The William Tell Overture* and buying what became the best selling Hartland western figures.

HOLSTER SET
1950's

This jeweled and studded leather holster, packing a pair of Texan Junior cap guns, was a typical sight to see strapped around the waists of mini-marshals. Many of these rigs had to be "surgically removed" just prior to teenage.

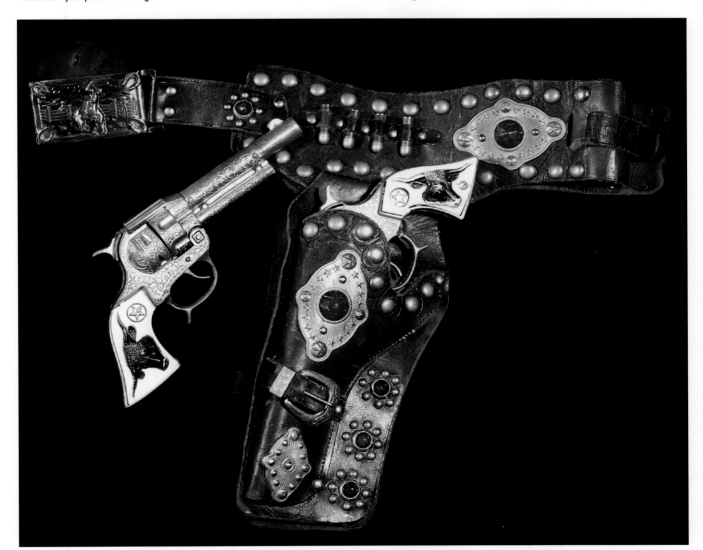

LAST CHANCE JOE
Rempl Plastics 1950's

There's gold in them thar squeak toys. A typical old prospector was personified by the makers of Froggy the Gremlin.

SHOOTIN' SHELL FANNER
Mattel 1958

"Fan 'em or Trigger 'em" but regardless of how you shoot Mattel's replica Colt 45, the cylinder revolves the next brass cartridge into the firing chamber, the hammer clicks down, and BANG! Out springs a plastic bullet tip zinging through the air. These smoking cap pistols were packaged individually or in holster sets made of "Genuine Top Grain Cowhide." The instructions offer some masculine advice to "Treat your gun the way a man treats a real gun. No man who depends on his gun will allow it to be banged up, dropped, thrown around, or left lying in the rain to rust."

CISCO KID AND PANCHO PUZZLE
Doubleday and Co. 1950

BONANZA
American Character and Palitoy 1966

In 1959, TV's *Bonanza* started a fire in Virginia City that blazed across the country and into the homes of 42 million viewers. For the next 14 years Bonanza pulsated as the hoof-beat of America becoming television's highest rated western. Finally in 1966, American Character re-created Cartwright family members Ben, Hoss, and Little Joe as eight-inch posable figures along with an outlaw dressed in black. Their horses came complete with harnessing and feature "action hoofs" which means there's a ball bearing for range riding the great indoors.

The set's toy highlight is the Bonanza 4-in-1 wagon. It comes with over 70 pieces including wagon parts, tools, and cooking supplies. The idea is to convert its design from a ranch wagon to an ore wagon, covered wagon, or for the Old West version of *Meals On Wheels*, the chuck wagon.

What few people realize is that after the Cartwrights galloped from sea to shining sea, they kept right on going with Bonanza being broadcast in 60 different countries. In conjunction with the popularity of the program, England's Palitoy simultaneously produced an identical toyline to American Character's with the exception of Little Joe. Palitoy's likeness of Michael Landon bears a much closer resemblance to the actor than the American version and is the one pictured here. Another difference was Palitoy hand-painting trim on a number of accessories. Adding natural colors enhance the guns, tools, and lariats but appear overdone on vests, saddles, and harnessing. With both the domestic and foreign toylines originally selling each character and horse separately, rounding up the entire set today seems to require the help of a pop-culture posse.

TEXAS STAR RANGER SPURS
Leslie Henry Co. 1950's

In the 50's at F.W. Woolworth, a retail price of 49 cents bought a pair of "Official Cowboy spurs that jangle."

GREENIE STIK-M-CAPS
Mattel 1958

Kids got a bang out of these glorious green sheets dotted with 120 circular self-sticking caps. The Greenies were designed to fit conveniently on the bullets and breeches of nearly every shootin' iron around.

FORT APACHE
Marx 1953

Indian attack! It seems the frontiersmen and women have just met the unwelcoming committee. Saturated with sagas of the Old West, kids whooped and hollered for miniature stockades. In response, toymaker Louis Marx supplied them with Fort Apache! The demand for this garrison was so great, it became the single most popular playset of all time with production runs spanning several decades. Along the way, the set evolved with changes to the figures, accessory inclusions, and the stockade itself. One character version worth blowing the bugle for was an official licensed Rin Tin Tin Fort Apache with figures of Lt. Rip Masters, Rusty, and namesake Rin Tin Tin reporting for duty.

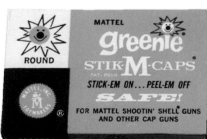

BUCKLE GUN SHOOTIN' SHELL
Mattel 1958

Hold your pants on! Here comes the "Look Maw, no-hands firing derringer," Mattel's Shootin' Shell Buckle Gun. Now I've seen everything. This comes in handy whenever a bad guy gets the jump on you, takes away your pistol, and has you reaching for the sky. Simply stick out your belly and an 1867 Remington model derringer pops out of your belt buckle to fire a plastic bullet at your captor. In case you feel a cough or sneeze coming on, a safety switch prevents unintentional firing. The deluxe version included a tooled western belt that I wish still fit my grown-up waist size.

THE RIFLEMAN
Hartland Plastics 1959

The Rifleman's horse appears courtesy of Bob and Barb Pierce

HOPALONG CASSIDY GAME
Milton Bradley 1950

HOPALONG CASSIDY
Ideal 1950's

For such a big western star it's surprising how small in the saddle Ideal made these toy Hopalong Cassidy figures. He has a moveable arm and removable hat which brings him up to all of 4 inches in height. Hoppy's horse Topper measures in at 5 inches high and has key chain reins. Together, wouldn't they look great galloping across ... a birthday cake?

THE RIFLEMAN FLIP SPECIAL
Hubley 1959

Here it is, the most famous western character rifle to fire from the television screen, The Rifleman's Flip Special. Each week, the hoop firing lever action was put to good use in the opening credits and later to resolve each episode. Only now with Hubley's cap firing replica it was our turn to flip the favorite firearm in hand upholding the law with some pretend "lead pumping."

The Rifleman Flip Special appears courtesy of Chuck Eckles

PALADIN
Hartland Plastics 1958

PALADIN HOLSTER SET
Halco 1958

In a word, what do you call a gun-slinger all dressed in black standing in the noonday sun? ... Hot! Yet as a kid, buckling up that same western star's black toy holster with Chess knight logos what do you become? ... Cool! But just in case someone in your gang remains unconvinced of your fame by association, whip out the card, *Have Gun Will Travel*. Then in anticipation of the card's future collectible value ... ask for it back.

Paladin holster appears courtesy of Chuck Eckles

HAVE GUN WILL TRAVEL GAME
Parker Brothers 1959

LARIAT
Ideal 1950's

Lariat appears courtesy of Rick Rubis

BAT MASTERSON CANE
Carnell 1959

Actor Gene Barry's suave portrayal of this western gambler would be incomplete without the signature gentleman's cane so often used as a billy club. A wood and plastic version was available with costumes or separately as a toy.

OUTDRAW THE OUTLAW
Mattel 1959

When kids with itchy trigger fingers began threatening to fill friends and family full of lead, it was wise for parents to buy Mattel's Fast Draw Target, Outdraw the Outlaw. The idea was to have a shoot-out with this false-front fiend of the frontier using Mattel's Shootin' Shell Fanner. The villain's firing speed could be adjusted to give beginners a fair chance, but hopefully, with a little practice, they'd soon be doing some mighty fancy shootin'.

ANNIE OAKLEY GAME
Milton Bradley 1955

ANNIE OAKLEY
Hartland Plastics 1950's

Opening credits of TV's *Annie Oakley* portrayed by Gail Davis began with "hard ridin' and straight shootin'." The courageous cowgirl gallops by, standing in the saddle, and shoots a bullseye through the 9 of Spades. It's no wonder, she became spokes-person for *Wonder Bread* whose ingredients promised "to build strong bodies 12 ways."

BEST OF THE WEST
Marx 1965

Going west with the idea of action dolls, Marx produced an original toyline called "The Best of the West." It began with Johnny and Jane West, their horses Thunderbolt and Flame, and Indian friend Chief Cherokee. Eventually, the West frontier family expanded to include son Jamie and daughter Josie, Pancho Pony, and dogs Flick and Flack. A plastic buckboard and covered wagon were made to scale delivering supplies to Johnny West's chipboard homestead, the Circle X Ranch. Each doll could be posed "1001 different ways" and came with enough outfits, equipment, and camp gear to stock a General Store. By 1970, the toyline turned pages of history adding General Custer and Geronimo to the popular series.

ROY ROGERS STAGECOACH
Ideal 1956

    Initially one of Roy's hired-hands resembling Gabby Hayes was given stagecoach driving duty before taking on the chore himself in later production. This was another toy in Ideal's fix-it and take-apart series which came with accessories including a rifle, whip, trunk and gold box.

ROY ROGERS CHUCK WAGON
Ideal 1956

    "Here they are! America's most beloved, most famous western television and movie stars!" Roy Rogers, Dale Evans, Pat Brady, and Bullet have brought "a wagon-full of Do-It-Yourself Fun!" including 60 pieces to take apart, put together, and play with. At chow time, a table folds down from the Chuckwagon to set up pots, pans, and "brew coffee." Finally, a noisy Nellybelle flaps its hood as pushed along the floor bringing Pat Brady in time for some grub.

IN DAYS OF OLDE

Take me back to those days of olde when men were knights and women were damsels in distress. Men got to clank around in skin pinching armor, heating up in the sun like *Spam*-in-a-can trying to impress fair ladies with pointy hats. People often spoke the word "Ye" and writers were further indulged as to add an extra "E" to words like toyes and giftes. Speaking of which, can you imagine how thrilled medieval Boyes and Girles would have been having toyes such as these to playe?

ARMORED KNIGHTS CONTEST SET
Sears Exclusive 1965

Taking the remote controller, aim the Silver knight's lance for the Black knight's shield. Likewise, a friend operating the opposing steed is steering a lance towards yours. The horses clip clop a simulated galloping sound with their heads bobbing up and down. The first to direct a hit, springs their opponent from saddle to floor. A pair of black and silver knight attendants accompany each champion with swords and shields held aloft as if cheering the contest. After Sears' exclusive distribution expired, the set was sold without the attendants, packaged in identical box art replacing only the Sears markings with the toy's manufacturer, Tomy. The duel shown is taking place before King Arthur's Castle which was produced in 1970 by Big toy company.

BEAR CLAW DAGGER AND SHEATH
Ohio Art 1950's

What respectable sword and sandal flick would be complete without someone flipping one of these through the air or to attach threatening notes on doors.

NOBLE KNIGHTS
Marx 1968

The whole idea of suiting up in armor is a thrill to kids of all ages. First you are practically indestructible, mighty deeds can be carried out in a shroud of mystery, and you get to clank around impressing the girls looking like a medieval robot. Marx went overboard creating this series of 12 inch posable knights with 20 pieces of snap on armor and 17 weapons including a sword and shield, mace, halberd, a firing crossbow, and even a 16th century pistol! (Who knew?) The instructions provide an interesting account of knight and armor developments from the 10th to 16th century. Horses were sold separately equipped with 17 pieces of armor and "hoof

hidden wheels that roll into battle." The series included the Silver Knight, Sir Stuart with horse Valor, and the Gold Knight, Sir Gordon with horse Bravo.

Marx also produced this set in England re-naming the knights, re-spelling the word "armour," and most significantly, adding Sir Cedric, the Black Knight!

By 1970, a Giant Medieval Castle was available made of chipboard and designed to fold up into a carrying case. Although massive with 3 square feet of interior courtyard, it still is under scale for the 12 inch Knights and Vikings.

COMBAT KNIGHT
Andy Gard 1950's

Stronger than dirt? Here's the closest thing to *Ajax* Laundry Detergent's white knight. Take the battery operated wired remote control to sharpen your knight's jousting skills scooping a pair of rings from a roly-poly practice pole.

Robin Hood

ROBIN HOOD CASTLE PLAY SET
Marx 1957

Robin Hood puts the jolly back in Olde England as his band of Merry Men make away with some loot being chased by the Sheriff of Nottingham in pursuit! Character figures Little John, Maid Marion, and Friar Tuck also help re-live the medieval tradition of knights, castles, and itchy clothing. And if you want more, there's a working draw-bridge, firing catapults, a regal buck, and a quiver of arrows that can actually stick into an archery target. The reason for so many features? ... the more the merrier.

Richard Greene, star of television's Robin Hood

THE BOLD VIKINGS
Marx 1970

Following Marx's impressive Noble Knights series inevitably came those horned hat heroic Bold Vikings! Each is outfitted with 5 pieces of simulated fur and leather clothing plus helmets. The weapons include a knife, sword and shield, working bow with arrows, an ax, and a spear. Marx Vikings pair up as the blonde Brave Erik and a bearded Odin the Chieftain. Their "Mighty Viking Horse" was available in brown or buff featuring head and neck movement and hoof wheels. Com-pared to the armored chargers, the Viking horse was skimpily clad with a mere bridle and reins, blanket, saddle, and stirrups. As with the knight's instructions, a brief history is included placing these warriors native to Norway between 800 and 1100 AD while further documenting their lifestyles and accomplishments. The dragon prowed Viking Long Ship in the background was produced by Renwal in the 1950's. (See The Fleet Elite)

SIR GALAHAD
Ideal 1950's

This toy representation of King Arthur's round table champion was irresistible to Ye olde children of the1950's. At the time, Ideal was known for selling take-apart vehicle toys and recommended the same theme of play, equipping Sir Galahad with 22 pieces including a sword, battle ax, and armor for his charger.

THE FLEET ELITE

In the 50's, a child's life was well defined. If given a toy boat we knew exactly what to do with it. Stick it in water. Mom and Dad were only too glad to provide us with these so we would stop sticking everything else in water. Available at the time were a fine assortment of sea craft to float in our tubs or play pools. Many of these were working theme boats approximately a foot long and came equipped with appropriate accessories. Then life changed. In the 60's, toy companies tried a gimmick of producing gigantic plastic battery boats loaded with function features for use on game room floors. Remco's shipyards first took the lead but later sailed against Ideal and Deluxe Reading before the fad faded.

Dry docked in huge boxes marked "Not a Water Toy" this unique series usually represented various warships. Their proud owners took them on Saturday afternoon voyages with anxious neighborhood crew kids. Together they huddled around these behemoth boats manipulating endless arrays of armament, plane launchers, blinking lights, alert buzzers, and bells. In its heyday, this fleet of non-floating fun was the envy of kiddie captains and admired by admirals.

Remco prototypes highlight Big Caesar's bow and stern detail.

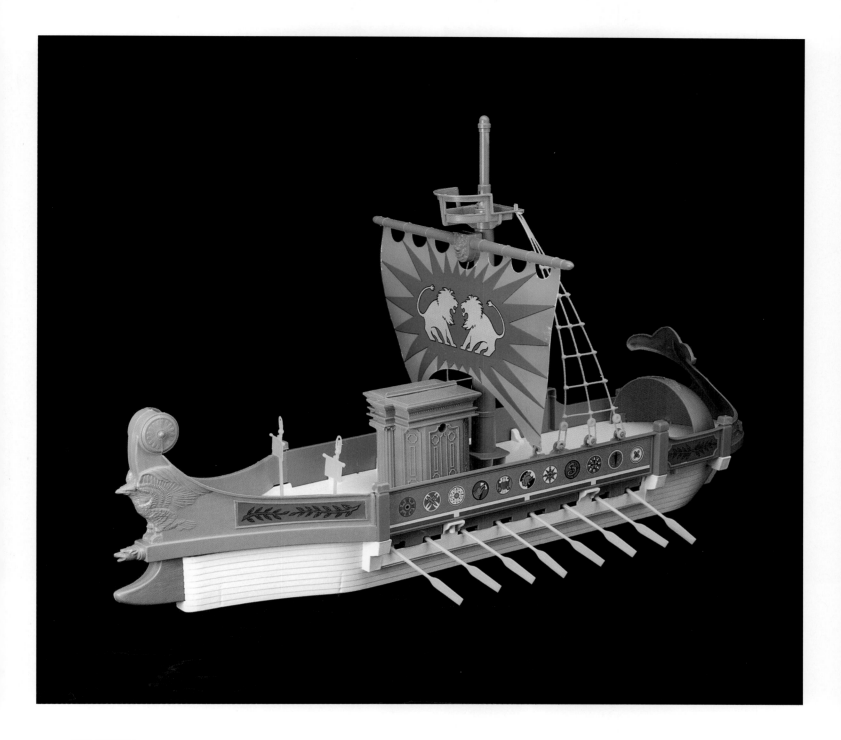

BIG CAESAR
Remco 1963

The fact that Big Caesar, a 2 1/2 foot Roman warship lacked actual seaworthiness was of no concern to little landlubbers in 1963. They simply set sail and hit the switch. With modern day batteries doing the work of galley slaves, rows of oars dip into an imaginary ocean causing the ship to lunge ahead. The vessel's glide fades until the oars swing back to repeat the cycle creating a realistic illusion of ancient sea travel. Then, as if this glorious galleon weren't enough, Remco also threw in two armies with firing catapults, chariots and a cardboard castle for good measure. A manually operated 1 1/2 foot version called Gallant Gladiator was also produced at a dinghy budget with ten yellow soldiers and one catapult. For those who see Big Caesar gliding through childhood dreams, bringing it back from the depths of several decades is pure treasure.

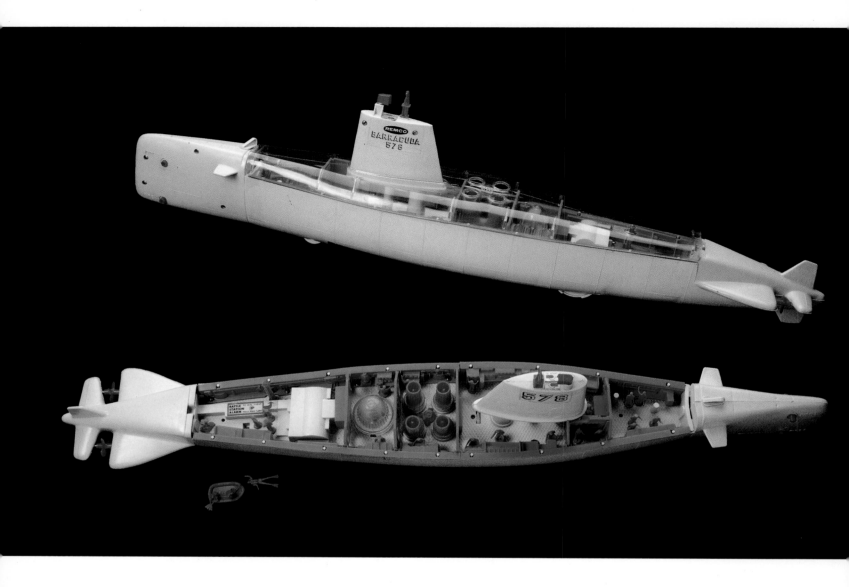

THE BARRACUDA ATOMIC SUB
Remco 1962

Remco's Barracuda was one lean mean submarine that more than lived up to the formidable reputation of its namesake. Red Alert! The nuclear reactor is blinking. Fire all four torpedoes off the prow! Okay, now let's get ready in case we need those nukes. Sir, the nukes are already firing. What? I gave no order for their launch! Well sir, they kind of go off automatically whenever the ship is proceeding forward. Really? How do we stop them? Sir, this continues until we either run out of nuclear warheads or stop proceeding forward. What is this a toy submarine? Yes sir. We are equipped with a 25 man crew including several frogmen and a raft. We have twin spinning propellers and a rod which can either activate a throbbing alert bell or be set to "run silent." Ensign you're color is looking all blue, better report to sick bay. Aye Aye sir.

FIGHTING LADY
Remco 1960

Don't get her mad. Or at least not the kid at the helm, because sometime when you least expect it, Fighting Lady may be rolled into position, her battery powered main gun rotated towards you and KER-BLAM! a plastic shell zings past your eye brow. Little ash can depth charges fling! Don't worry about the amphibious craft hanging from the ship's stern. It's harmless. However, there is one last feature, unintended as a weapon yet the most lethal, the reconnaissance plane. A rear deck launcher can zap that bugger across a room at a speed that would have Chuck Yeager ducking his head. Thank God for its rubber tipped nose or I might not be here today!

US NAVY FROGMAN
Remco 1961

Tooter Turtle's friend Mr. Wizard should remind this frogman "Be just what you is ... not what you is not." In water, (lots of water) posable arms and flippers guide or steer the US Navy Frogman powered by a single battery operated spinning propeller. Next, yank him out of the "sea" and water spews out from holes riddled throughout his body. Now get this, you're supposed to plop an army helmet on his head and hand him an M-1 carbine and a walkie-talkie. This sight probably has the enemy scratching their heads. "Who's this blue guy full of holes walking around wearing flippers?"

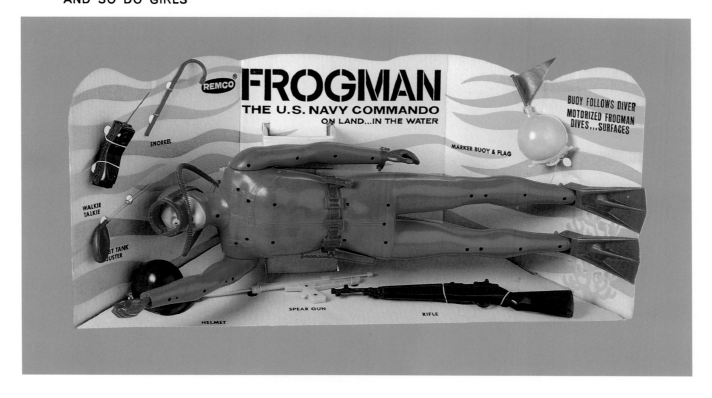

SHOWBOAT
Remco 1962

    Possibly taking inspiration from the Judy Garland/Mickey Rooney films, one of Remco's designers must have shrieked "Say! I've got a swell idea ... We'll put on a show!" "Great, let's draft a sketch of an elegant Mississippi riverboat and include a script book for kids to perform plays on the boat's stage. How about punch-out characters and scenery for Pinocchio, Cinderella, Heidi, and The Wizard of Oz? Then we can add some battery functions like forward motion and a spinning paddle wheel or maybe some footlights, right?" Wrong. Apparently someone cut the budget before any battery features could be added. Yet even without lights or motorized movement the curtain calls were made because as you know ... the show must go on!

FIREBOAT
Ideal 1955

Ideal's Fireboat satisfies one of childhood's most basic motor skill needs, squirting water. On land, unwind the anchor and tug the boat along on wheels as a floor toy. Then plop it in the nearest sea and turn the hand crank. An alert siren begins to whine while at the same time engaging water cannons that "really squirt water!" Finally, a pair of life preservers and life boat finish equipping the ship for whatever hub-bub hits the tub. Now there's just one more thing bothering me about this whole concept. Did you ever wonder if a fireboat hosing down another ship used so much water extinguishing a fire that they ended up sinking it? Sounds like a good plot for a 3 Stooges film.

PIRATE SHIP
Ideal 1953

What better to buccaneer with than Ideal's Pirate ship? As the skull and crossbones flap above furled sails, turn the pilot wheel toward another adventure. Just don't turn your back on the cut-throat crew who may be in the mood for someone to walk the plank. In the event a ship worthy of plundering sails by, the Ideal pirates are ready to fire cannon balls from the deck or swashbuckle with standard issue of cutlass and dagger, or flintlock pistols. Ideal also produced a friendlier pink variation of this ship with Mickey Mouse as its captain and a crew of other Walt Disney cartoon characters.

THE PHANTOM RAIDER
Ideal 1964

There's something fishy going on. What appears to be an ordinary freighter has more to it than first meets the eye. At first, the only action stirring is the ship rocking from side to side as if riding the rhythm of waves. More simulated sea travel includes forward or reverse propulsion and manual steering. A bell continually sounds as it voyages along the floor. Although these functions add to the toy's realism, it doesn't seem to be a military threat ... Not yet.

Other toy warships cruise past the Phantom while kiddie captains ignore what appears to be an insignificant craft. Then a strange transformation takes place. The Phantom Raider extends its mid-section to reveal three hidden bays. Before you can say, "Battle Stations!" the hatches pop open, ready to fire a salvo of torpedoes, missiles, and depth charges at an unsuspecting enemy. About the only thing that could save you now is having a secret stowaway on board the Raider named Bond ... James Bond.

MIGHTY MATILDA
Remco 1963

As toy war at sea waged on, Remco's shipyards launched Mighty Matilda to recreate the Navy's incredible "beehive at sea," the aircraft carrier. By engaging switches, Matilda gets mighty with forward drive, a rotating radar pole, signal alarm bell, and motorized elevators. An over-scaled 100 man crew has little to do except stand in line taking turns at the ship's single rocket launcher.

Then finally, the mission is at hand. Position squadrons of aircraft for take-off. One by one, press the runway button to send 12 pursuit planes, 4 helicopters, and 9 bombers into the next room yonder.

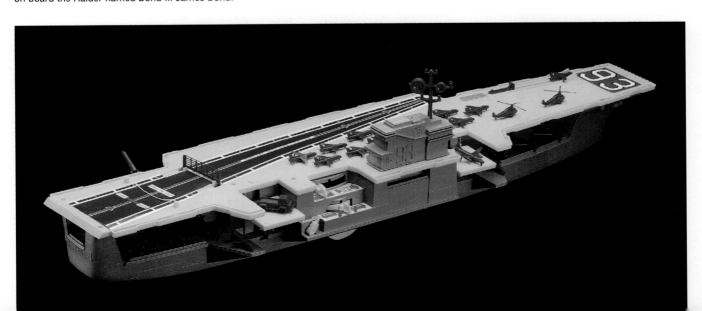

BATTLEWAGON
Deluxe Reading 1963

Need a Flagship for your fleet of unfloatable boats? How about one with enough firepower to blast all the others down the drain? Put on your admiral's cap and step aboard the USS Battlewagon.

As the ship is launched, we scout for a worthy opponent to test our weaponry. Since no other sea craft are cruising the floor today, we'll set course for the Basset Hound sleeping by the heat register. This is going to be touchy. If all goes well, Battlewagon will slip in, blast the beast with all rockets and torpedoes, and quickly exit unscathed. It sounds good on paper, however problem number one involves the sneaking up part. An alarm bell continually rings as the boat begins its forward sea rocking motion.

To our surprise, the alert signal is being ignored by the "enemy dog" which is now in range of the forward guns. Only the upper two cannons shoot with the lower three for decorative intimidation. Additional special effects include a bag of powder to create a realistic puff of smoke from the barrel. Ready? FIRE! A salvo of two missiles arc and begin descending. It's going to be close ... No! Missed by a nose! The creature yawns with canine indifference.

At this point Battlewagon's side torpedoes are broadside, nearly point blank with the snoozing seadog's shoulder. The danger of a ricochet exists, but it's a chance we'll have to take. FIRE ONE! FIRE TWO! FIRE THREE! Our worst fears are realized as the torpedoes bounce off Bowzer's back. The captain reports that without reverse or steering the ship is unable to make a second pass. We must abort the mission. First one concern remains: to retrieve all projectiles before they are mistaken for dog yummy treats. Reconnaissance planes are launched to locate the darts, enabling landing craft to recover.

WARNING! The hypothetical demonstration in Battlewagon's review was merely a descriptive dramatization. Boys and girls reading this at home are not encouraged to attack pets or siblings with their toys. However, it does offer an explanation as to why so few projectile firing toys are found today complete with accessories. Some fell prey to missile chomping dogs or teething babies. Though in most cases, Moms were famous for confiscating any weapons or ammo which could prevent Junior from growing up to finish medical school.

VIKING SHIP
Renwal 1950's

Unshakable images of Vikings sailing aboard their legendary Long Ships will forever conjure those ancient sea warriors with a lust for plunderous exploits. At least that's the way Kirk Douglas played them in the movies. The way some lucky kids played them in the 50's was with Renwal's Viking Ship. The forbidding Dragon's head prow originally intended to ward off foes now has the opposite reaction befriending fans of the adventurous craft. Its hull is designed to serve in sea or on land with wheels underneath. Actually, the toy acts out a much better floor show than it does afloat. When pushed along the ground, Viking figures are activated to row the oars in unison. The disadvantage of the figures permanently positioned for seafaring duty is offset with a swiveling catapult which fires plastic balls. If you are still stuck for the want and need of combat figures carrying broad swords and battle-axes, may I suggest a name mightier than Thor, more powerful than Odin. The name you seek is Marx!

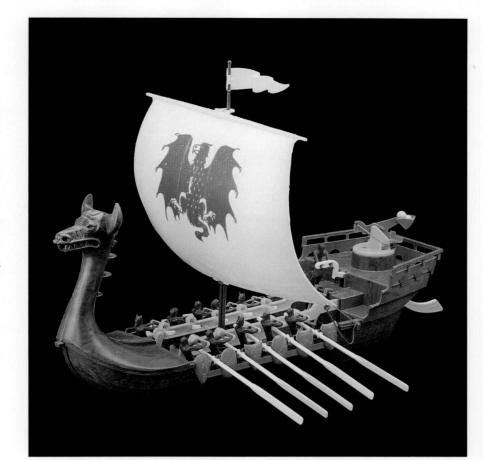

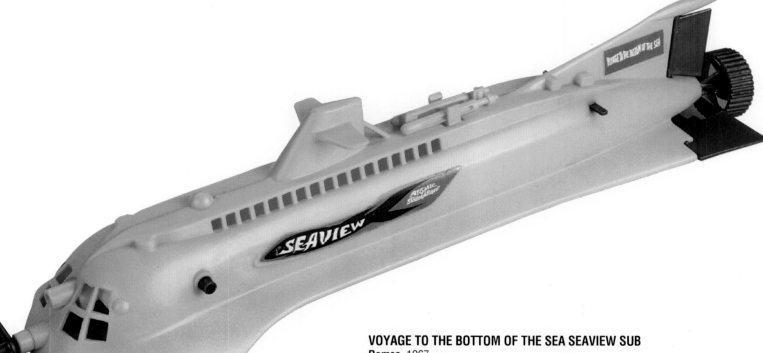

VOYAGE TO THE BOTTOM OF THE SEA SEAVIEW SUB
Remco 1967

No matter the danger, be it enemy depth charges or giant squid, television's Seaview would shake, rattle, and roll through treacherous waters to safety under the courageous command of Admiral Nelson and Captain Crane. Once Remco's official sub was launched, anxious kids could now put their own Seaview into welcome waters to re-enact the show's jilted adventures. The toy was packaged with various accessories as per catalog listings. One popular Sub set came with a pair of divers, a floating whale, and a sea monster (a refugee from an earlier Hamilton's Invaders set.) Regardless of the accessories, the Seaview was designed either for use on land or in water. Using "elastic motor propulsion" (a rubber band) the propeller spun while rear rudders controlled raising, or submerging, and turns. A pair of torpedoes could be spring launched at the monster. family and friends. pets ... in fact, it's hard to "fathom" all the possibilities.

KEYS TO YOUR FUTURE

Ask today's bright grade school gang what America stands for and they'll proudly sing out, "Freedom!" Then walk past a couple buildings to High School and ask a group of sixteen year olds what "freedom" means. They'll reach into their pockets and purses and hold up car keys!

Since the invention of the automobile, America has evolved into a "drive-through society." Reflecting the world on wheels, various versions of kiddie cars have long been a staple of the toy industry. So, why don't we peel out, lay some tire, and pop a wheelie with some of these classics. We'll skip the traffic light fire drills and forgo mooning "pressed ham."

SUPERIOR GAS STATION
T. Cohn 1961

Back then these tin buildings miniaturized a faded phenomenon known as the "service station." Uniformed attendants would be in garage bays changing tires, brake pads, or adjusting carburetors until someone pulls in double-dinging the black hose. This universal signal alerted mechanics to drop everything and go pump gas. The dispensing of fuel in those days was a ritualistic ceremony which included offering to check under the hood, tire pressure, and cleaning the windows. As late as the 1960's, service stations conducted business like a circus sideshow competing with giveaway premiums. To encourage fill-ups, motorist lures ranged from inflatable dinosaurs to stick on horse shoes, plush tiger tails, red antenna balls, dishes and glassware.

T. Cohn's colorful rendition is equipped with both an elevator and ramp to access upper floor parking. Besides the pumps, accessories, and service crew, an unusual feature provides storage of customer's cars on shelves until needed. Countless variations of tin gas stations prevailed throughout the 50's to serve pre-licensed playees and are today a prized find for any automotive enthusiasts.

BILLY AND RUTH AMERICA'S FAMOUS TOY CHILDREN 1951

PLAYMOBILE DASH
Deluxe Reading 1961

What more needs to be said beyond toy box hype which boasts, "The Most Exciting Toy Ever Made!" It's almost as if Deluxe Reading found the snazziest console Detroit had to offer, ripped it out of the car, wired up D-cells, and plopped it in a toy box. Amidst this chromed creation's gaggle of dummy gauges are battery operated features including a running motor, working windshield wipers, lighted turn signals, and a beeping horn.

Of course everyone knows the first step to driving a car. Ask Dad for the keys and hit him up for some gas money. Maybe he'll forget it runs on batteries and slip you a couple bucks.

CRUSADER 101
Deluxe Reading 1964

If bigger is better, Crusader 101 is best! How big? A warning slip inside the box pleads for kids not to sit on the incredible 2 1/2 foot cruiser for a ride. As far as the design goes, it seemed standard practice for Deluxe to borrow features from actual makes and models without actually becoming any one in particular. Then for no explainable reason a statuesque driver was placed behind the wheel who can best be described as ... tan. Aside from the boring "Mr. Tanman," the rest of the car is one slick ride cast in typical Deluxe red plastic with black and white interior and gobs of dazzling chrome trim. If you need to change a tire, there's a spare with tools in the trunk. Now, to take Crusader out for a spin, a battery operated wired remote controller does the trick. As an oversized convertible this would have been the ultimate for a trip to the drive-in movie or for curb service burgers, but when it's time to parallel park this boat ... better call an admiral.

CRYSTAL STARFLITE
Ideal 1956

Here's a cool cruise for the curious. Ideal took the "visible model" idea and made it into one of their "fix-it, take-apart" toys. This time the 20 inch long clear plastic body is just one of 130 dismantling parts. Batteries work the head and tail lights while other manual features include steering, working pistons, fan blade and belt, and a radiator that can be filled and drained. One thing to remember when you have a see-through car ... it's not recommended for dates at the drive-in.

BIG RED
Marx 1965

Doesn't this look just like one of those super huge impossible model kits that you had to actually read the directions to build? Except it isn't. Just slide this baby out of the box, stick on the roll bar and horn, pop in some batteries and away it goes. I really mean "away it goes" because there isn't any wire or remote control to keep it from eventually crashing into the wall! Don't just sit there reading this! Go grab the steering wheel and turn those mag wheels for a circle pattern, or down shift into low gear, or put it in reverse to retrieve the toy! See, this is what happens when someone just reads the first line and then hits the road.

Big Red also comes with a jack for mini-mechanics to check out that heavy metal cast chassis. Sorry girls, box hype suggests "Be the first to join the trend ... with a boy's best friend."

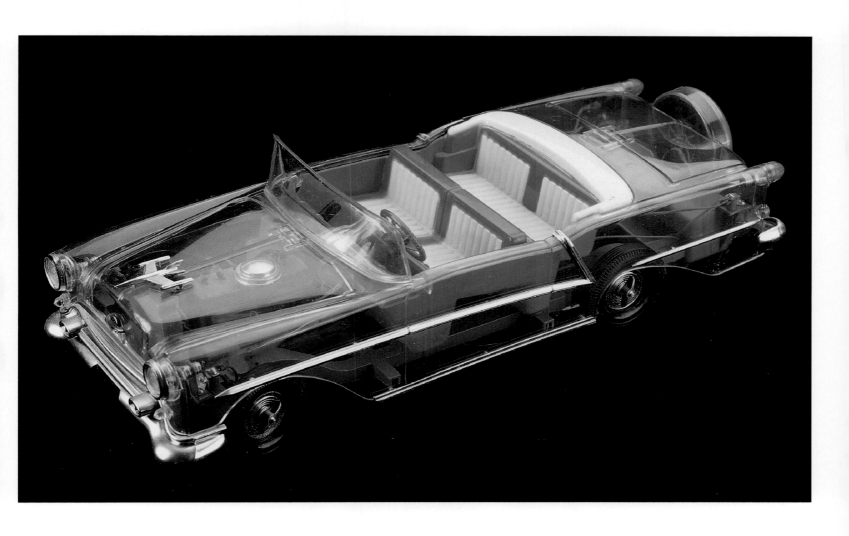

BIG RED

THE BEVERLY HILLBILLIES CAR
Ideal 1963

Come and listen to a story 'bout a man named Jed. In the midst of the Hillbillies heyday, Ideal must have set for a spell pondering just what in tarnation they should produce as a toy from the hit television series. The answer was as obvious as a tick on a flea's ear ... the family's beat-up truck! Thinking big, they started out 22 inches from tailgate to grill and then loaded it up with a rickety bench, table, chair, ladder, barrel, basin, and a spare tire. A number of other hillbilly belongings include a rifle, saw, stool, and wash board, but only the frying pan can actually be held by Granny. The rest is excess baggage.

Jethro Bodine gets to drive with Uncle Jed riding shotgun. Behind them Ellie Mae shares a bench with the bumpkin's bloodhound. Then on top of the heap, sits a determined Granny. Now to send the Clampetts on their way, simply point the truck in the direction of Beverly Hills Californy, git out front and crank 'er up. Away they go with a shimmy and a shake.

FLYING DUTCHMAN
Remco 1962

A year after Remco's U-control drive Shark Racer took-off for the toy market, the same means of propulsion was U-tilized in this antique car version. Additional features include a working headlight and a man made of chipboard who appears to have popped from a frame tray puzzle into the driver's seat.

MR. MAGOO CAR
Hubley 1961

Not that I'd recommend driving this across telephone wires or through a pig-pen screaming "Road hog!" but, by George, try popping a pair of batteries in to watch wacky wheels drive our nearly blind buddy in circles on a rickety rat-a-tat ride. "Oh Magoo, you've done it again."

DISNEY PARADE ROADSTER
Marx 1950's

Disney Parade Roadster appears courtesy of Rick Rubis

TRIK TRAK
Transogram 1964

The Trick is in the track! Toss some straights and curves on the floor leading the Trik Trak Racer as you go from room to room. Design a different layout every time with disjointed track and punch-out paper scenery. Several vehicle variations were produced including a Crazee Cycle set.

MR. KELLY'S CAR WASH
Remco 1963

Tired of reading "Clean Me" in finger dust writing on your car? So, take it down to Mr.Kelly's Car Wash. After hooking on to a cable, this battery powered automatic system pulls your vehicle along from start to finish. With clear plastic window walls observe the car beginning the wet-down phase, then continue through cleaning and drying rollers. Outside, touch it off with little towels, sponges, and real wax, all included.

Mr.Kelly's Car Wash appears courtesy of Jeff Silna

MOVIELAND DRIVE-IN THEATER
Remco 1959

   Hop in the car. It's almost dusk. We're going to the Drive-in! Kids playing with Remco's re-creation of this American phenomenon didn't have to wipe off windshields or raid the snack bar to get ready (maybe the refrigerator). They did get to "drive-in" a half dozen tin toy cars, pull up to the speaker poles and roll the features. If they were too antsy to wait for night, a black cardboard hood enclosed the toy for daytime darkness. Double feature slide shows presented captioned adventures of Mighty Mouse and Farmer Alfalfa, Dinky Duck and Captain Kangaroo, Heckle & Jeckle and *Have Gun Will Travel*. Even miniature marquee cards were provided to advertise the current features. Of course if this toy were made today, an extra card would have to be included ... "Flea Market Sunday!"

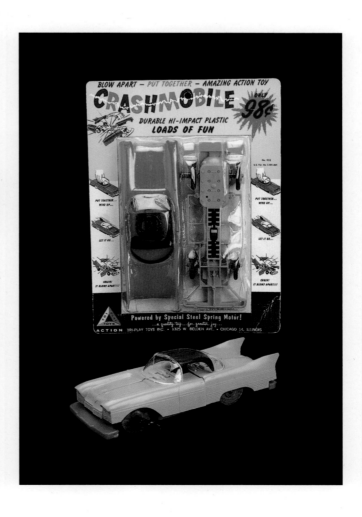

CRASHMOBILE
Tri-Play Toys 1960

   Finally, a toy you're not only allowed but SUPPOSED to smash-up. First, snap on fenders, hood, trunk, and roof. Then wind-up the spring-loaded axle, aim for the wall and let 'er rip! Upon impact, the car is triggered to EXPLODE! Tri-play's crashing car variations include an early model with windshield, one without, an antique variation, and a secret agent car. Several other of their wind-up wonders from the blister pack toy rack include Break-A-Bull, Boomerang Tank, and Jousting Knights.

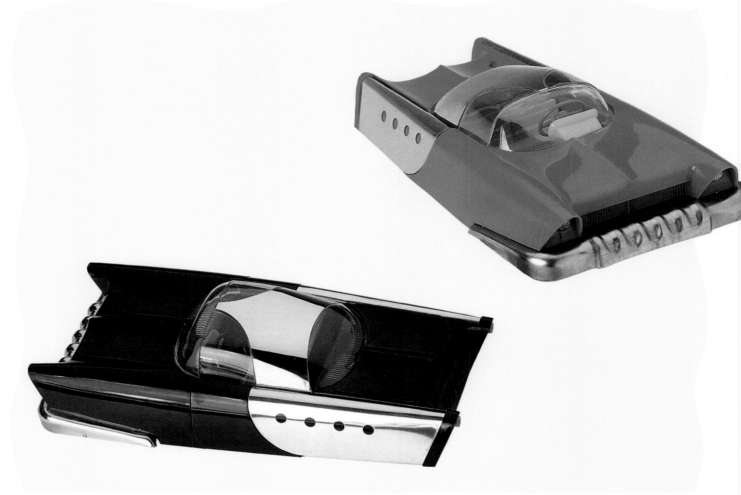

DREAM CAR
Mattel 1953

Described on the toy box as "futuristic," Mattel thought they were previewing the motoring market yet to come. Unfortunately, Detroit somehow skipped over the spacey designs imagined by Mattel's toyline. At least kids were able to enjoy these super cool hardtop convertibles. To go convertible, gently squeeze in at the sides of the "bomber bubble" to remove. Then go for a ride with several pushes to rev up the friction engine and let 'er go! It's one smooth cruise ... like a dream.

XP 600
Ideal 1953

The XP 600 "Fix -it car of tomorrow" looks ready to blast off. Besides using a 7 piece roadside repair kit, other battery features include a honking horn and working headlights. A cardboard sheet of license plates was provided for each of the 48 states which as Think-A-Tron should know is minus Alaska and Hawaii.

SHARK RACER
Remco 1961

    With the design of an Indy 500 speeder, Remco's Shark Racer is ready to revolve the scenery of any room. Once revved up, a pull of the U-Control handle sends the Shark on its course. The controller steers the car in circles of either a five, ten, or fifteen foot radius or can be set for straight drive. The Shark's claim to fame actually comes from Remco's own promotional literature. In 1961 they included a miniature comic book catalog inside packaging of their toyline. The plot involves taking a little boy named Jimmy on a cruise in the Shark through Remco Land to visit various toy areas. Shouldn't they have driven something else with a trunk?

V-RROOM GUIDE WHIP RACER
Mattel 1963

    Hear that V-RROOM real motor roar! The idea is, you control the car tethered to a wooden guide stick and it "whips" around your race course. The Racers are sleek and the motor really growls. So much so, that the ad slogan was altered to "You can tell it's Mattel, it sounds swell!"

HOT ROD HELMET
Ideal 1960

V-RROOM!

While America's space program was in its infancy, a courageous country gasped as each development slowly transformed science fiction into science fact. Whenever space shots occurred, classrooms across the nation became "junior mission controls" with television sets wheeled in to witness the historic event. As students from across the hall filed in to share the TV, the room's capacity filled to "standing room only." Finally, all the fidgeting and chatter was shushed to silence when the time reached T-minus one minute and counting. All eyes fixed on the dart-shaped rocket as if it were the Times Square ball at 11:59 New Years Eve. When the countdown reached T-minus 10 seconds, kids whispered along with the technicians at Cape Canaveral, 3-2-1-Zero, Blastoff!

Meanwhile, the world's toy manufacturers were simultaneously launched into orbit. After having filled the 50's with an infinite variety of robots, rockets, and ray guns, present day reality could now be added for even more futuristic fun.

ASTRONAUT HELMET
Ideal 1961

It didn't matter what else you were wearing. Anyone plopping one of these on their head became, for that moment in time, a real-live astronaut. Ideal's space helmets were equipped with tinted visors and a kazoo type speaker which turned every kid into "Buzz."

THE JETSON'S FUN PAD GAME
Milton Bradley 1963

Surprisingly, Fun Pad is one of few toys made from the successful prime time animated TV series, *The Jetsons*. A 3-D plastic space dome similar to the hovering structures in their cartoons is balanced on a Fun Pad pole. After taking turns loading up the landing pads with dozens of space cars the object of the game becomes a challenge not to tip it all over.

LITTLE HONEYMOON
Ideal 1965

Dick Tracy presents "My new grand-daughter, Little Honey Moon." Proud parents Junior and Moonmaid are shown on the cover opposite Grand Dad with their birth announcement. Little Honey Moon is a 14 inch doll that cries and comes with "Special bow-tied Moon Hair" sticking out of her removable space helmet. To commemorate her birth, an 11 panel comic strip replays the event around box flaps.

ATOMIC CAPE CANAVERAL PLAY SET
Marx 1960

A record included with Marx's Atomic Cape Canaveral sets the mood with a narrator reading instructions for Junior Missile Officers. Eventually, he announces "This is a live launch. All unauthorized personnel will now leave the ready room and lift-off area." In the background at Central Control actual technicians can be heard "Roger Wilco-ing" each other amid beeps and effects. This is followed by a countdown, blastoff, and tracking officer who states "The missile is on course. The satellite is in orbit." If all that doesn't get you ready to spring off a dozen rockets and saucers, you must be from another planet!

OPERATION X-500
Deluxe Reading 1960

It's the ultimate space base. Deluxe earns its name providing two enormous eye-popping toys in one. A tail finned rocket gantry prepares to launch man into space (or ceiling) along with several other X-500 capsules containing a dog, monkey, and satellite. The silo is also equipped with both an elevator and crane which function manually by hand crank. Base Defense, looking more like Base Offense, is armed with an arsenal of missile launchers and features a chrome instrument control panel loaded with buttons, dials, levers, and switches, some of which spin, ring bells, or light up. In addition, there's a helicopter and a score of crewmen, technicians, and guards for more hands-on fun. Sitting at the controls of Operation X-500, one is sure to accomplish a mission into the imagination.

ROCKET BASE USA
Deluxe Reading 1960's

Deluxe Reading offered Rocket Base USA as both a companion set to Operation X-500, and as a lower budget alternative to their feature toy. The main attraction for expanding Operation X-500 was Rocket Base's mobile rocket transport which is space lingo for "truck." This nifty flatbed carries three boosters, (Tell the driver to go easy around the bend.) and is chrome trimmed including a futuristic antenna which seems to resemble Nabisco's logo. Otherwise it stands on its own as a fully functional missile launching rocket base with a dozen technicians and if you believe box hype, is "the most amazing toy ever made!"

It's true. In the early 60's before Deluxe Reading became Topper Toys, they sold their wares exclusively in supermarkets across the nation. A row of store displays occupied the air space just above and beyond our reach, glistening in glory.

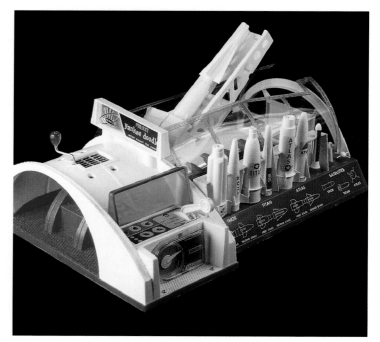

THE GOLDEN SONIC
Tigrett 1957

Tigrett describes the Golden Sonic as "The unique toy unlike any other in this world." "The remote control toy completely guided anywhere by sound!" "It responds to every command of a special super sonic whistle." "The magic radar screen hunts for the sound wave and actually follows it." "The unique toy that looks and performs like a million dollars!" Could P.T. Barnum possibly have said it better?

PROJECT YANKEE DOODLE
Remco 1959

A Top Secret rocket silo hides its launcher under tinted bay doors until blastoff. Then ... it's time! Sound the alert signal! Countdown lever engaged. As the needle approaches zero, a launcher arises from the station pushing through tinted bay doors, 3-2-1-0 Whap! A two-stage Thor missile zaps across the room. The Pentagon would be proud.

MOON SHIP
Knickerbocker 1950's

Jeepers creepers, what a peeper! In fact, the sole passenger of Knickerbocker's Moon Ship is a great big eye ball which can be rotated by the operator's "magic wand." Once the eye is looking in a desired direction, engaging Moon Ship into forward propulsion will send it along the surface towards that destination. The craft is also equipped with a pair of firing rockets which should be carefully aimed so as not to "put someone's eye out."

ELECTRONIC Z-MAN, THE BRAIN
Z-Line 1956

Our over-grown space cadet Z-Man, who appears jammed into a kiddie car, has just taken off. He's on a programmed course to blast a pair of targets with electronic guided missiles. As he whips around a couple of pylons, tail lights alternately flash. The targets are nearly in range. Now to see if your coordinates are properly zeroed in. Here goes ... ZZZZZAP! Whoa! a direct hit! Wait he's turning, another ZZZZZAP! He shoots and scores!

The secret to programming Z-Man lies hidden underneath his silver helmet. By shifting it forward, a printed circuit board is revealed. This is Z-Man's "Brain Central" where course plotting is set using a series of 20 buttons. Until you get the hang of it, several basic maneuvers are charted to help Junior cadets get started sending Z-Man on his next mission.

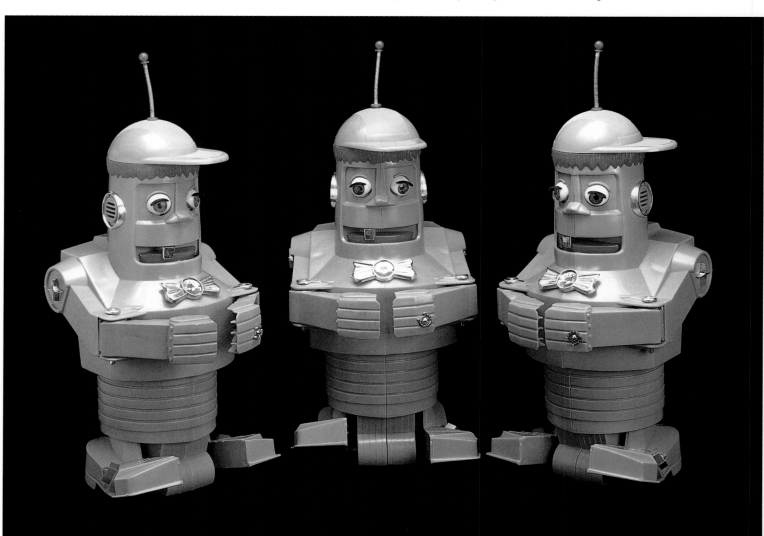

ELECTRONIC COUNTDOWN
Ideal 1959

It's 1959. As we click on TV and twist the tuner dial to a favorite cartoon program there seems to be a live rocket launch being televised, 3-2-1-0 Blastoff! No wait, it's just a commercial for Ideal's Electronic Countdown. Now here comes the toy version. "Hey, my toy doesn't do all that!"

It was and still is common for prototypes to be used in early promotion. Slight differences in design were hardly noticed but when a toy's functions are enhanced to the point that the rocket silo emits steam and launches with the sound of dynamite exploding, one might be disappointed to learn that these features were just studio effects. Otherwise, Countdown certainly lives up to the price of admission with lever activated battery powered controls, a rocket transport cart, revolving radar dish, and launching silo that sends your missiles "from the Moon to Mars."

YAKKITY YOB
Eldon 1961

The name of a robot often refers to its functions or purpose. In this case, Yob is boy spelled backwards and Yakkity refers to his squeaky speech. Like a puppet, the little red-haired robot is operated by placing a hand inside the toy to activate moving arms, blinking eyes, and his namesake - mouth moving squeak speech. Next, take him for a walk by pushing his clickitty feet along the floor. Yakkity Yob comes equipped with a baseball cap which can be rotated and is topped with a springy antenna. He also wears a glad/sad bowtie which permanently registers in the glad range, and a horseshoe ring for luck.

Yakkity Yob
...made by ELDON

ELECTRIC ROBOT AND SON
Marx 1955

Here comes cover boy! It's the Electric Robot from December of 1955's *Time* magazine. Actually, the robot was one of many Marx toys pictured with their world famous toymaker Louis Marx in an article featuring his company and career. Actually, this toy's significance goes beyond some good PR on the newsstand. He was the first plastic toy robot born from his alma mater and better still, a great toy.

With battery powers abounding Electric Robot was aptly named. At the flick of a switch bulgy eye bulbs light and away he goes,

propelled across the floor turning his blockhead grin from side to side. A buzzer in back could be pressed for some dotting and dashing of the inescapable Morse Code. Other built-in doohickey highlights were a chest drawer with miniature tools, an antenna to raise or lower from his head, and twisty knob arm movement. Later a nifty afterthought was the inclusion of a diapered baby known as "Son." Our photo portrays a family reunion of both color variations.

MYSTERY SPACE SHIP
Marx 1962

Long before the word Gyro became popularized as a Greek treat at shopping mall food courts, it had scientific significance. To Junior astronauts of the 50's and 60's, gyroscopic power was a means of propulsion put to good use by the Marx Toy company with their miraculous Mystery Space Ship. The toy's gyro energy was unleashed by hand cranking a rotor to the speed of 4000 revolutions per minute thereby creating a centrifical force with "power to perform strange feats." Instructions suggest 52 different tricks to try while educating

readers with the explanation of gyroscopic principles. In order to earn several licenses you are quizzed as to the knowledge and understanding of terms like spin and torque vectors, gimbal axis terminology, and precession principles. Upon mastering gyro theory you are bestowed with the Interplanetary Navigator's License. Then again, there were kids of the age or mentality (and we know who we are) that simply set up those neat little Moon men, shot off a few rockets, cranked up the ship, and said "Gee, this sure feels funny!"

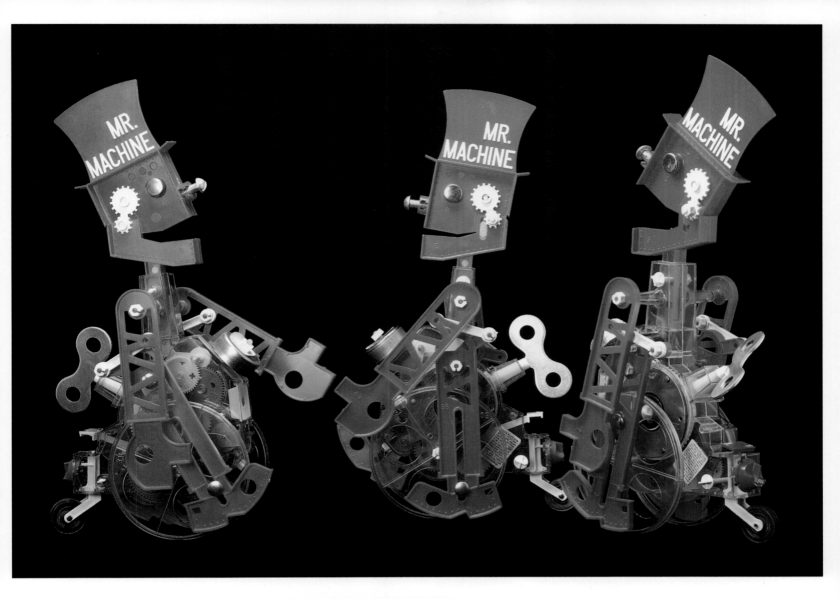

MR. MACHINE

Put-Together Assembly

MR. MACHINE
Ideal 1960

"Here he comes! Here he comes! Greatest toy you've ever seen and his name is Mr. Machine!" Originally, Mr. Machine came ready to roll but could be disassembled and re-assembled following a 7 step diagram with a little white wrench and a series of 44 numbered levers, gears, and thing-a-ma-bobs. Turning a giant wind-key sent him into an arm swinging, head tilting, leg marching promenade while ringing a bell and occasionally squawking from a belly box. Re-enforcing his fame, a board game was produced in 1961 involving little plastic Mr. Machines on a race to the robot factory. Their moves were selected by rolling a disc through the body of another specially designed Mr. Machine. For a time, the popularity of Ideal's miraculous mechanical man earned him the position of spokesperson for the company's other toy commercials. Following a promo, an animated Mr. Machine became instantly assembled and proclaimed, "It's a wonderful toy ... it's Ideal!"

SATELLITE CAR
Ideal 1956

Ideal took Outer Space to the open highway with this satellite launching vehicle. Revolve the bubble domed passenger pod from side to side, adjust radar screen, angle back the launcher and crank it up. After loading one of the flying saucers in place we are ready to hit the firing button. SWOOOOOSH! The saucer spins into a blur whispering its way to the ceiling or into your next door neighbor's yard.

LOST IN SPACE ROBOT
Remco 1966

"Danger! Danger!" Somewhere lost in space lies the galaxy's most famous, most quoted unknown robot. Officially referred to on television as "the robot," our cantankerous chum is now affectionately called "The Lost in Space Robot." Remco's toy version traveled along the floor, could be set for turns, had manually moveable arms, chest canister, and blinked lights in his chest panel. Don't ask him do anything else or he is liable to respond: "That does not compute."

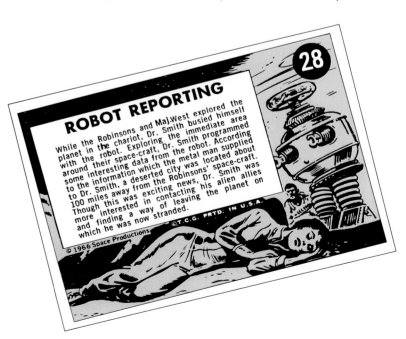

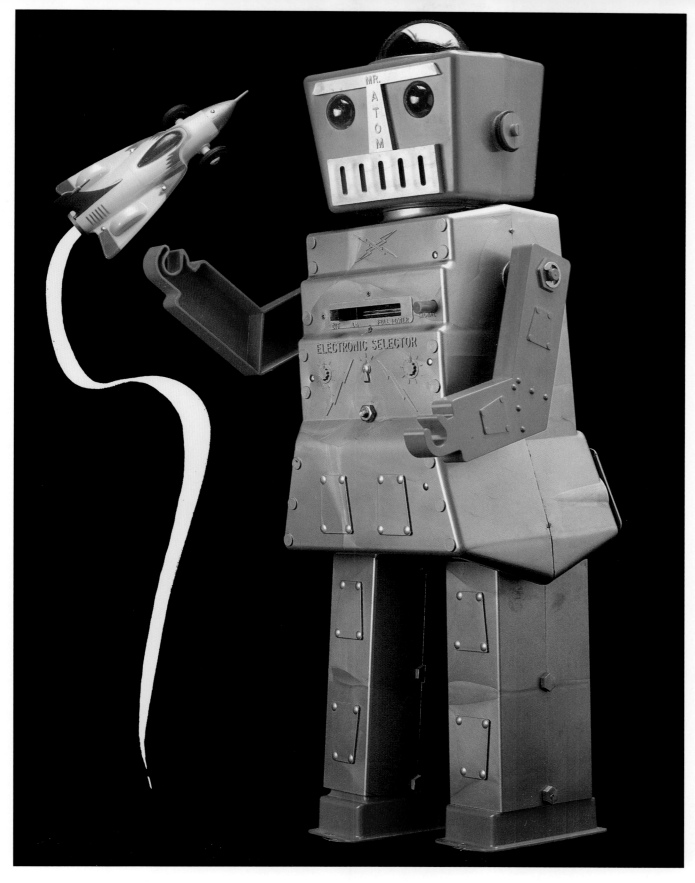

TURBO - JET CAR
Ideal 1956

Attach the Super Sonic Turbo-Jet Car to its tin litho launching platform and energize by cranking up the spring mechanism. At the push of a button, it "zooms away with screeching sound."

MR. ATOM
Advance Toy and Novelty 1950's

Watch out for this lanky blockhead. After hitting the switch, off he goes with manic arm and leg movement, his head turning from side to side. You can further aggravate him by pressing a button which both lights the head while sounding a buzzer. Or let's make Mr. Atom go totally berserk in his full power setting. In this mode the speed is accelerated causing the arm and legs to become particularly frantic while the light and buzzer continually throb as if the robot is about to explode. If there was such a thing as a TNR index "toy nightmare rating," no doubt Mr. Atom would be in the top ten of his era.

OPERATION MOON BASE PLAY SET
Marx 1962

Prior to astronaut Neil Armstrong's " ... giant leap for Mankind," most of our knowledge about the Moon came from cartoons and comics. From this unquestionable source, we learned about the Lunar terrain consisting of green cheese and that Moon Men would also be green. (I guess from eating the cheese). Further proof of these theories came in toy form such as Marx's Operation Moon Base. This playset featured a three dimensional greenish chunk of Moon complete with craters, a Lunar mountain range, and 64 other pieces of *Crayola* colored space stuff. Some of the Moon goodies included were an exploding mountain, a suction cup spider rocket, a missile firing Moon Ship, saucer and rocket launchers, an underground silo, orbiting space station, Lunar terrain vehicles, 6 pleasantly peculiar Moon men, and some really big tires!

MR. POTATO HEAD ON THE MOON
Hasbro 1968

One year before America's Apollo 11 mission successfully put man on the Moon it appears the place was previously visited by Mr. Potato Head aboard the Cucumber 1. Being one of the late 60's "headless sets" we re-enacted his adventures with earlier "hollow heads" since Mom was expecting company and wouldn't let us raid the 'frige.

ROBERT THE ROBOT
Ideal 1954

In classic blockhead style, the Ideal Toy Corporation produced the world's first plastic toy mechanical man, Robert the Robot. It seems the toy has proven the test of time re-establishing his irresistible popularity with today's collectors. His original demand according to Robert Malone's *The Robot Book* resulted in 500,000 kids adopting the toy in the 1950's.

Functionally, the robot's basic feature is to scoot along the floor via a cable drive mechanism hand-cranked by his operator. The early version also had a few extra spiffy effects like swinging arms and a removable "belly button" housing a mini-metal hammer, screw driver, and pliers. Although Robert's head would always light, the early style with clear bubble eyes and antenna seem to create a more stunning display. However, Robert's show stopping performance was always to hear his voice record inside. Turning a crank in back plays the message "I am Robert Robot, mechanical man. Drive me and steer me where ever you can."

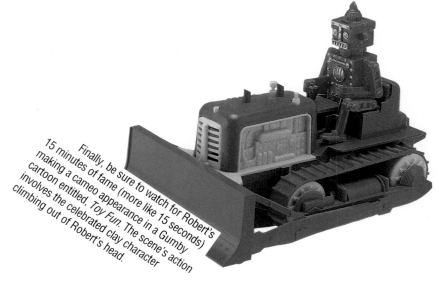

Finally, be sure to watch for Robert's 15 minutes of fame (more like 15 seconds) making a cameo appearance in a Gumby cartoon entitled, *Toy Fun.* The scene's action involves the celebrated clay character climbing out of Robert's head.

ROBERT THE ROBOT TRACTOR
Ideal 1955

This tiny version of Robert carries on the family tradition as an illuminated blockhead, however this time, twisting a knob atop the head sheds light on the subject. The tractor itself has some slick features with an adjustable blade and rolling rubber treads operated by the same cable drive controller as the original Robert the Robot toy. As the tractor changes direction, little robotic arms shift levers as if he's the one doing the work.

ASTRO BASE
Ideal 1960

Ideal's television commercial transforms a child's play environment with sand, craters, and mountains into a planet from outer space. The young space cadet so indulged as to demonstrate the nearly two foot tall Astro Base also gets to wear a Colonel McCauley astronaut helmet. At first glance you're already hooked before he's even touched the toy! Special effects begin with movement of scanners rotating in the dome. An Astro Scope follows, revolving landscape in a viewer. Then to engage the toy's feature highlight, switch the power panel's space lock which raises a bay door while extending a hoist. Then try lowering the astronaut to the remote controlled scout car below. Adjust the radar screen for steering and take off for a lunar cruise. Finally, to complete your mission of interplanetary play, pre-set timed firing cams launch a pair of missiles "out of this world."

BIG LOO

BIG LOO
Marx 1963

Maybe Big Loo does have the kind of face only a mother or a generation of kids from the 60's could love. Personally, I take a look at his bullet-shaped head, the toothy grin, those blinking red eyes and I'm taken back to 1963 when our "friend from the Moon" entered the household. We were all over this gooney guy either trying to decipher the wobbly speech from his voice crank, or testing his array of communication and defense systems. There were darts shooting out the chest, balls from an arm, and a multi-finned rocket sprang from a foot launcher. He even had a water squirting syringe lodged in his belly button. Besides the projectiles, other non-aggressive features to fidget with included a built-in scanner scope, a whistle, bell, compass, and a clicker to send Morse Code messages. Since few of us were fluent with Morse code, a sticker chart was provided in back.

To get around, Big Loo's foot base contains wheels enabling him to be pushed or pulled places. Further exercising of our Lunar Looney involves bending at the waist while using a triggered arm with a pincher claw to pick up objects. One peculiar accessory which can be picked up and tossed in this manner is a light green something which resembles a futuristic microphone but is supposed to be a Moon grenade. I guess you or Loo throw it, it lands somewhere, and then you pretend it goes "Boom!" Well, maybe that feature is kind of dippy but considering the robot and everything included selling for an original retail of $10 it was the biggest and best money could buy. In fact, he still is. The only thing I'd rather have than Big Loo would be ... two Big Loos!

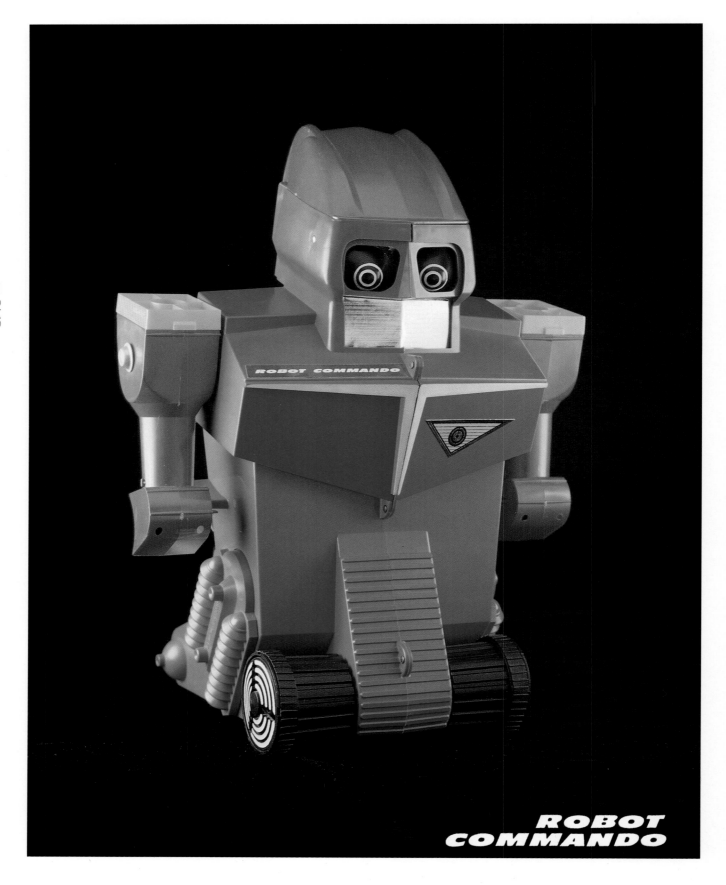

ROBOT
COMMANDO

ROBOT COMMANDO
Ideal 1961

Rising up from the city skyline comes the one and only Robot Commando! With his menacing war-like design, kids were only too eager to get at the controls. A wired remote controller which was originally voice activated sent Robot Commando on his missions of mayhem which included pitching missile balls from arms and raising the top of his head to fire a rocket out of a hidden silo. This all occurs while black "jelly bean" eyes rotate a hypnotic stare, and mysterious squeaks emit from within. Directionally, Robot Commando moves forwards and turns. There is no need for reverse since he never retreats. National Defense at the Pentagon would be glad to know this attack robot is one of ours.

STEVE ZODIAC'S
FIREBALL XL5

FIREBALL XL5 SPACE SHIP
Multiple 1964

I suppose the worst mission to send a crew of marionettes would be into a zero gravity environment like outer space. Undaunted, producer Gerry Anderson launched his tangled tales aboard the Fireball XL5 into sixties sci-fi history. Multiple's 20 inch toy version simplified matters with plastic figures of Steve Zodiac and friends Venus, Commander Zero, Professor Mathew Matic, Robert the Robot, and

Zoony the Lazoon to re-play the show's spazzy space adventures. Like on TV, the capsule XL5 Jr. detaches for quick planetary landings. Slide open the hatch to remove the pilots and their Jetmobiles. The Fireball itself is armed with spring loaded rockets which fire from opening bay doors. All this excitement and more was also available in Multiple's expanded playset, Fireball XL5 Space City.

SPACE RANGER Orbiting Space Ship
Marx 1962

"A real flying toy as seen on TV!" "You are the pilot. Fly it fast, slow, hover, take off and land by remote control." Or, my favorite ... "Bomb the exploding Space Saucer!" The ship really whips around, connected to an aluminum rod spanning about a six foot diameter. You also get a gaggle of astronauts, Moon Goons, an X-21 towered Earth Base, a Lunar Base, space mat, space car, parachute, and rocket launcher to perform your mysterious missions.

BIG MAX Electronic Conveyor
Remco 1958

Turn a knob and Big Max pivots into position. Now shift levers to control his every move. Once his "powerful magnetic hands" are lowered into a bin of metal discs, activate electro-magnetism. A dome atop his head lights to signal the power is engaged. Suddenly the discs are collected as if by magic. More lever shifting and Big Max deposits them on a revolving conveyor belt which in turn loads the discs into a pickup truck. Now after observing Big Max in action, another form of magnetism pulls in the next operator. "Hey it's my turn!"

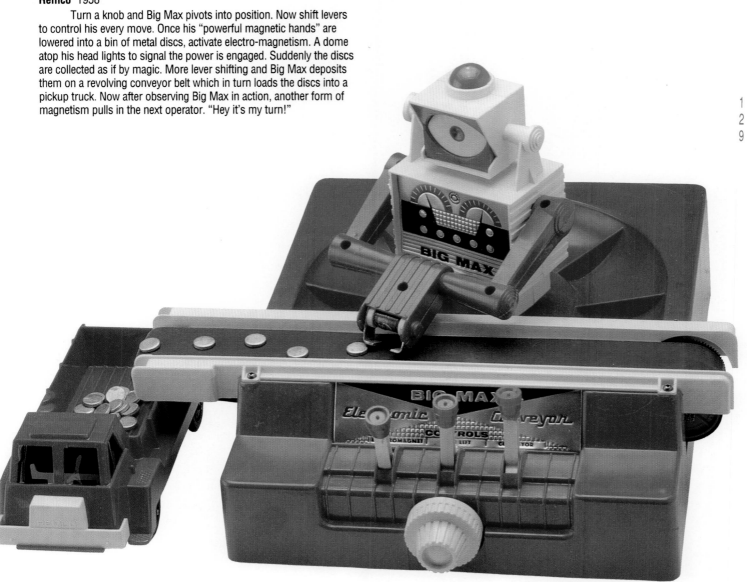

MENTOR
Hasbro 1961

Go ahead and challenge Hasbro's electronic wizard, Mentor, "The Man of Bronze who thinks for himself" by indicating lighted responses. The game includes a series of 8 theme cards such as Treasure Hunt, Danger Land, or Return to Earth and promises "hours of fascinating fun."

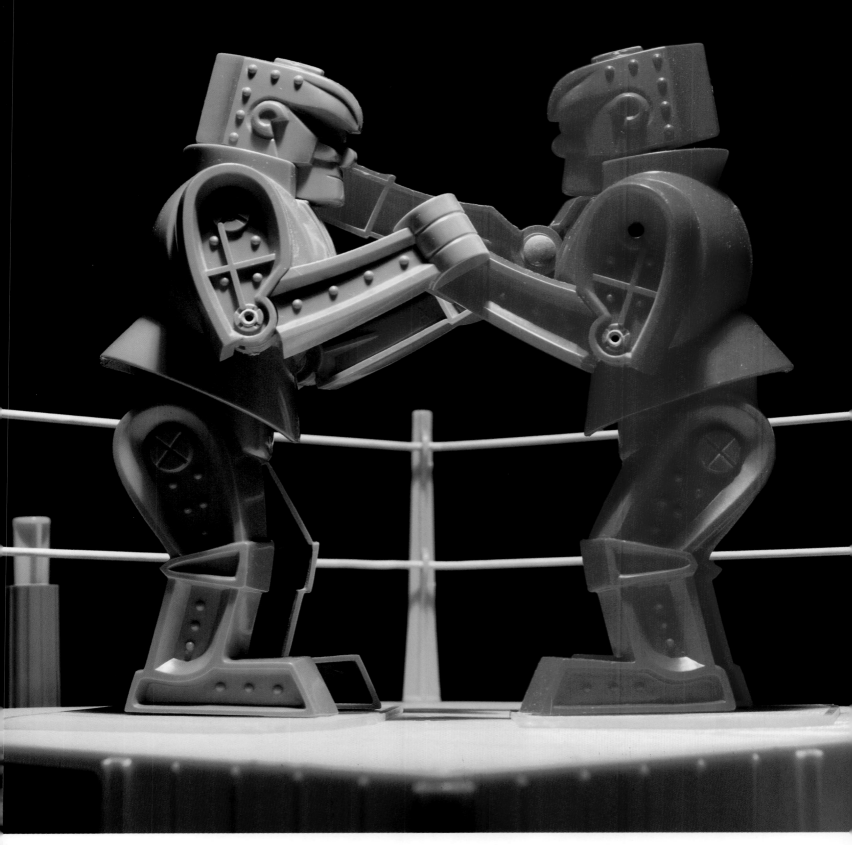

SCORE CARD
FEATURE BOUT
OF THE EVENING
THE
ROCK'EM
SOCK'EM
ROBOTS

The beautiful
BLUE BOMBER
from Saultarus II

The
rollicking
RED ROCKER
pride of Umgluk

round	1	2	3	4	5	6	7	8	9	10	totals
RED ROCKER											
BLUE BOMBER											

Knocking head loose
wins and ends round . . . .

TH'
WINNAH!

ROCK 'EM SOCK 'EM ROBOTS
Marx 1962

Box hype welcomes you to the "Heavyweight Championship of the Universe! ... with the world's only boxing Robots in a fight to the finish! From Soltarus II weighing 375 pounds it's the rollicking Red Rocker versus the pride of Umgluck, weighing in at 382 pounds the beautiful ... Blue Bomber!" Beautiful? Who else but Marvin Glass and Associates would plop a couple robots in a boxing ring to duke it out? I don't think there's a person alive in the free world who doesn't know what happens next. You Rock 'em. You Sock'em and BIZZZZZZZ! knock your opponent's block off!

MAJOR MATT MASON
Mattel 1967

Mattel's Man in Space, Major Matt Mason, using the "Bendy-Principle" of inner wired posable figures, took aspiring astrokids imaginations into the galaxies. Coming from the makers of Barbie, the toyline's marketing program followed suit selling each figure and accessory separately with the exception of certain deluxe space station sets. Differing from most science fiction themes, Major Matt Mason and friends were conducting a non-violent exploratory mission. Any equipment firing projectiles or resembling a weapon were actually sending space probes, search globes, space sensors, or taking readings with exploratory laser beams according to package descriptions. Even aliens like super human spaceman Capt. Laser with flashing eyes, Calisto from Jupiter with transparent green brains, bug-eyed Scorpio, and Or the strange shaped visitor from Orion were all extra-friendly-terrestrials. The rest of the crew along with Major Matt Mason included Sgt. Storm in red space suit, Radiologist Doug Davis in orange, and Rocketry Specialist Jeff Long in blue.

TELEVISION SPACEMAN
Alps 1961

This must be the robot all the others gather around for entertainment. To start the show, insert an antenna inside his head and away he goes. In an arm and leg moving march, eyes rotate and a mouthful of colorful lights revolve in sync with a panoramic spacescape in the robot's belly.

SMOKING SPACEMAN
Line Mar 1960

Produced in Japan by a division of the Marx Toy Company, Smoking Spaceman captures the classic look of 1950's science fiction with his silver metallic blockhead expression. After taking the typical arm and leg synchronized stroll, he stops to blink red eye bulbs several times while emitting a stream of smoke from his chrome bumper mouth. To top it off, a red and blue revolving "barber shop cone" lights up inside Ol' Smoky's dome.

SUPERIOR SPACE PORT
T. Cohn 1952

It's an airport. It's a stockade. It's a gas station from space. Actually, the Superior Spaceport is a clever conglomeration which shares tin with each of those other three T. Cohn playsets. This bizarre creation continues its borrowing spree by commandeering Captain Video Space vehicles, launchers and Space creatures. Superior's Spacemen are a surrealistic squad of army figures molded in shiny metallic plastic and equipped for galactic adventure by virtue of having amber helmets plopped on their heads. Other spiffy outer space features include a tin litho firing cannon, a saucer launcher atop the tower, and a hand cranked viewing screen which rolls images of aliens and their attacking spaceships. One spazzy do-dad has to be the warning siren clicker which rivals the realism of a New Year's Eve noise maker, (which is probably what it is).

MR. MERCURY
Marx 1961

Although this metal man is operated by wired remote control, you'd swear orders are given from a control room inside his head. A close look reveals his face shield to be a window wall with what first seems like eyes actually turning out to be tiny technicians sitting in chairs. Mr. Mercury's motorized movements include walking with lighted chest panel and name plate, bending over, and arms that move in to grasp objects. The robot came either silver or gold with early production models receiving an extra overhead helmet light.

Mr. Mercury appears courtesy of Dick Brodeur

GI JOE SPACE CAPSULE
Hasbro 1966

Three, Two, One, Zero, Blastoff! Actual excerpts of John Glenn's historic Friendship 7 mission are heard on a record enclosed with Hasbro's GI Joe Space Capsule and astronaut suit. Other facts and information including the ship's orbit speed (17,500 mph) and re-entry heat shield temperature (3,000 degrees) are revealed. The toy itself is based on blueprints of NASA's Project Mercury, aside from a sliding hatch feature to provide access for play. Parents understood "loud and clear" that kids were "Go!" for this GI Joe.

FIELD HELICOPTER
Sears Exclusive 1967

This is just what every kid needs. Strap this on your back, reach for the lever, and get ready. Colorful lights blink, and an overhead propeller spins. Woody Allen tried one of these in his futuristic film, *Sleeper.* In the movie, to help the unit achieve lift-off, he runs downhill flapping his arms but never quite gets off the ground. Neither does the toy.

GI JOE
Hasbro 1964

"GI Joe, GI Joe, Fighting man from head to toe, on the land, in the sea, in the air." Those very words, sung to the tune of *As Those Caissons Go Rolling Along*, kicked off a television commercial campaign introducing little boys to a new action soldier who was about to take the country by storm. Hasbro's mission was an incredibly risky maneuver. Never before was a posable doll with accessory clothing ever handed to a boy by the toy industry. Some parents and retailers scoffed at the idea. Even a number of toy industry veterans warned of disaster. Undaunted, Hasbro proceeded with their plan. Military manuals of uniforms and weapons were consulted to insure authenticity of detail. According to Marvin Kaye's *A Toy Is Born*, television

commercials were launched as an advertising offensive resulting in the capture of approximately 35 to 40 million dollars in sales during the first two years of production. Toy war continued to wage on throughout the 60's with military vehicle additions to the line by both Hasbro and licensee Irwin Corp. Foreign soldiers also enlisted in Hasbro's service and even a nurse. By this time as you might have guessed, all the opponents and skeptics of the GI Joe toyline had either surrendered or were reported missing in action.

All GI Joe Action Soldiers, GI Joe Jeep, and GI Joe Space Capsule appear courtesy of Bob Pierce.

FROM THE HALLS OF MONTEZUMA

Only recently have platoons of 1960's toy veterans come out of the bunker to recover their former military toys. For some reason, people have always accepted toy weapons associated with either medieval times, the old west, cops and robbers, or ray guns from outer space. But the minute a single drop of olive drab paint touches a toy, someone always screams, "War Toys!"

I remember screaming too. Conducting backyard war games, as one of a nerdy bunch of kid commandos yelling and carrying on, teaming up against imaginary foes. "Come on men! Let 'em have it!" A salvo of pops, clicks, rat-a-tats, and cap bangs fill the air. Other times when we split up to oppose each other, a different noise filled the air ... the arguing of who got who. "I got you first!" "Did not!" "Did too!" "Oh yeah, well you were in the open and I was behind the tree." "So what. I got you when you stuck your head out!" "Did not!" "Did too!" The who-got-who honor system always had its heated moments of dispute.

Somehow maneuvers conveniently ended by lunch time with a rendezvous at central headquarters (a plastic pup tent). Toy guns were dropped in a pile. Play-ammo was scattered throughout the yard. Right now, our more immediate concerns were rations of cookies, candy, and potato chips. The two guys who were screaming at each other are now sharing swigs of cola from a sticky canteen. Dusty faces streaked with sweat laugh over the mornings calamities. Nobody died. Instead, we had the time of our lives.

TIGER JOE
Deluxe Reading 1961

Here it is, the biggest battery operated tank ever made, Tiger Joe! Maybe you didn't know tanks came with chrome trim. Well they do at Deluxe Reading. A wired walkie talkie acts as the remote controller which rolls the treads forward or reverse while a turret gunner rotates. Then pop some powder down the barrel before loading in the shells. Click back the lever and hit the firing button! In a puff of smoke, Tiger Joe zaps your target with amazing accuracy.

MIGHTY MO
Deluxe Reading 1964

What a name ... Mighty Mo! This was an army howitzer version of Remco's Johnny Reb Cannon. Both measure in around 2 feet and could easily whip a ball across a room, bounce off a wall, and come whizzing past your ear on ricochet. Mighty Mo had a huge breach loading shell which required ram-loading a ball on the tip. It could be fired either by push button or an optional wired remote cable controller at adjustable elevations.

JOHNNY SEVEN O.M.A.
Topper Toys 1964

The ultimate one man army, Johnny Seven O.M.A. is seven guns in one! Ready? Set? One! Grenade Launcher. Two! Armor Piercing Shells. Three! Anti-Tank Rocket. Four! Retractable Bi-Pod. Five! Repeating Rifle. Six! Automatic Pistol. Seven! Tommy Gun. Hey, wait a minute. What's with number four, Bi-Pod legs? They aren't a weapon feature. But if you check number two Topper does provide a pair of different styled armor piercing shells. So maybe it should be, Johnny Six and a Half! Even with this slight mis-count in firing chambers, the Johnny Seven O.M.A. blasted the toy market wide open. In Topper's 1965 catalog, an open letter from President Henry Orenstein accounts for over 1,600,000 of the Johnny Seven rifles selling in their initial year of production. In attaining this figure Topper topped all previous toy industry records for single year sales in the toy's price range.

JOHNNY SEVEN MICRO-HELMET
Topper Toys 1964

By "Micro" they are referring to a microphone built into the amber visor which connects to a wired walkie-talkie. Beware getting "hung up" in backyard jungles. Instead, forget the walkie-talkie. Just disconnect the wire, grab the Johnny Seven O.M.A. and let em' have it!

Johnny Seven

WHIRLY BIRD
Remco 1960

    For most people, a two foot copter spinning blades through the air generally leaves a lasting impression, (or a poke in the eye.) Whirly Bird probably did both for those who experienced this classic chopper. Other battery operated features included forward drive for on-ground movement and an operating winch. A platform could be raised or lowered by the winch to supply 25 troops with a tank, truck, jeep, and howitzer. On board, a hinged door opens to transport the army to their next field operation.

BULLDOG TANK
Remco 1958

    Remco's toughest tank, the Bulldog, is lever activated to climb over blocks, forts, sister's dolly, or stationary pets. A second lever fires plastic shells into battle while ejecting the brass cartridge. The same actions apply to the smaller Light Bulldog Tank.

REMCO MONKEY DIVISION INSIGNIA

1
4
2

LONG RANGE BAZOOKA
Remco 1960

Do you have a problem with annoying enemy tanks prowling through the neighborhood? Maybe there's a miniature army playset you'd like to set up and then obliterate with a few blasts. Whatever the occasion, simply pull the loading rings back to set for distance, take a gander through scope sights, and Fire away! "Harmless shells" are now whizzing their way to those we deem unfriendly.

COMBAT GAME
Ideal 1963

Our hero, Vic Morrow as Sergeant Chip Saunders, shows off his Thompson sub machine gun in front of Lieutenant Hanson played by Rick Jason. Give me Saunders, Kirby with his B.A.R., and the Combat theme song and we'll win any battle in 60 minutes (minus commercial time).

DEFENDER DAN
Deluxe Reading 1964

Now we're getting serious. Pull back the slide bolt to spring wind the firing mechanism. Move thumbs into position and let 'er rip. A cartridge belt feeds 24 plastic bullets past the firing gate whipping the shell out the barrel and kicking out the spent cartridge. Defender Dan also swivels and elevates so no one gets away.

GUERRILLA GUN SET
Mattel 1963

Mattel's camouflaged Thompson sub-machine gun with flip-up sights and adjustable cloth strap is just like the one Sergeant Saunders used on Combat, only better. It's yours. Slide bolt action either blasts caps or rat-a-tats when you're out of ammo. The set also features a camouflage poncho, Commando Beret and jungle knife.

MARINE RAIDER LONG RANGE MORTAR
Remco 1960

With fold-up bi-pod legs, Remco's mortar quickly collapsed for Saturday soldiers on the move. The trick is to lob your hollow shells towards a cardboard pillbox target considering your distance setting, angle of elevation, and hope to heck it isn't windy.

INTO THE WILD BLUE YONDER

STEVE CANYON HELMET
Ideal 1959
    Visor down. Snap on your kazoo speaker oxygen mask, and get ready for takeoff!

Based on aerodynamic design, it is supposedly impossible for a bumble bee to fly. It is further explained that the reason bumble bees do fly is because they lack this information.

People are in the opposite situation. We already know aircraft are capable of flight. We know there will be in-flight magazines, or in-flight movies, peanuts and soda pop, Salisbury Steak and Chicken Kiev. Undaunted, we buy a ticket and go. So why is it, once leveling off at 30,000 feet looking down at the tops of clouds, that it finally occurs to question the physics of the situation?

A much more sensible group of problem solvers, kids, already know their toy airplanes can't fly so it is up to them to achieve and maintain altitude, (three feet above shoe level).

FLYING BOXCAR aka GLOBEMASTER
Ideal 1956
    Ideal's Flying Boxcar caters to the military with vehicles and troops to go. But before coming in for a landing there had better be some clear floor space to set up all this neat Army stuff. Included are a Jeep, truck, firing cannon, launching missile, adjustable searchlight, radar, and a squad of little blue actionless soldiers either sitting, standing, or walking. Hey, couldn't someone have told Ideal "We're looking for a few good men?"

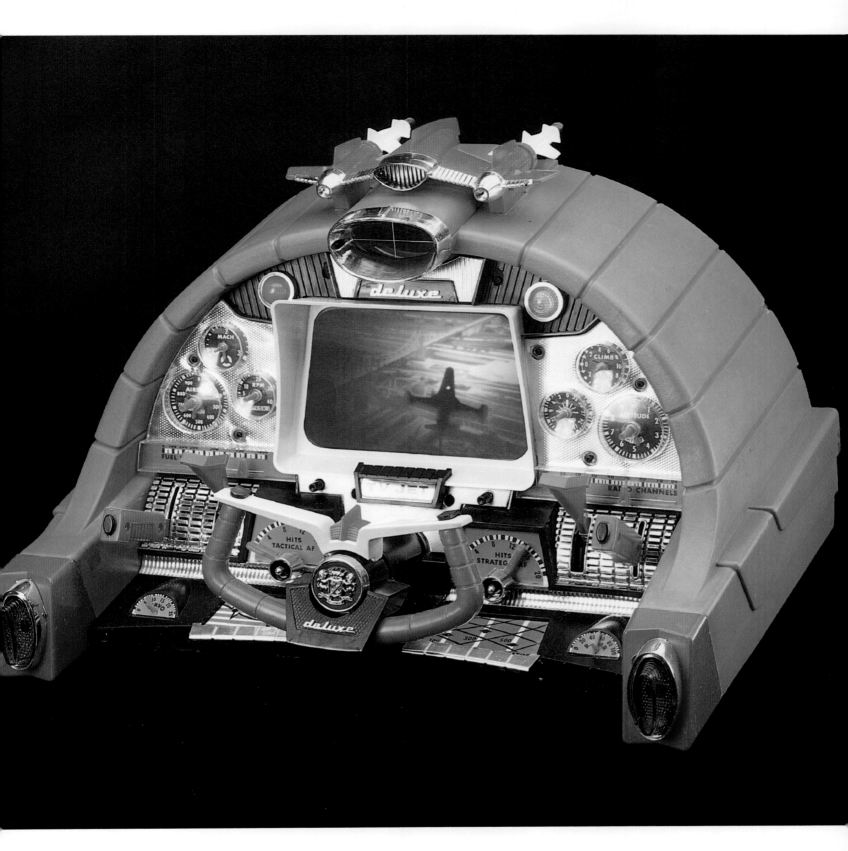

JIMMY JET
Deluxe Reading 1961

"Gee, it looks more like a rocket panel than a plane's." With Jimmy Jet, miniature minds went into a maximum daydream mode gawking at its chromed gauges and colored reflectors. Manual adjustment of those instruments set everything from rate of climb to radar. However, the toy's highlight is a motorized TV Jet screen which lights up and revolves an aerial view of various landscapes. A jet silhouette moves above the scenery guided by the toy's steering wheel. Now, spot your target and reach for firing levers to zap missile darts from atop the console. Mission accomplished. "Great! My turn!"

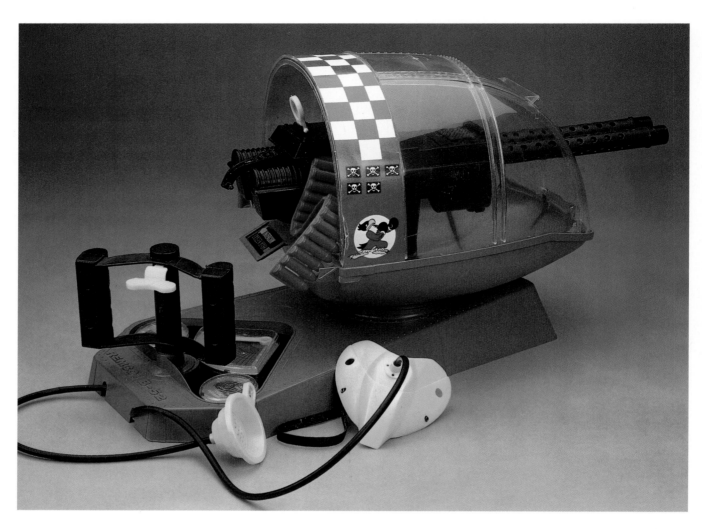

B-52 BALL TURRET
Remco 1961

In the 60's, Remco sent courageous kids and their imaginations to the skies dueling warbirds as gunner in a B-52 Ball Turret. First, strap on the oxygen mask and confirm your position over the intercom. A quick inspection of the gauges reveals the ammo level, rounds per second setting, and a turret traverse indicator. Readying for attack, lift the machine gun's cover to feed in the cartridge belt. Snap the cover down and raise the sights. Handle grips control both the pivoting of the of the ball turret as well as the firing button. Depressing the trigger lights a pair of gun barrels accompanied by rat-a-tat sounds while feeding the bullet belt through its chamber. Turning the canopy from side to side activates a radar scope which, at this moment, detects approaching aircraft. Here they come! Stay sharp!

ELECTRONIC FIGHTER JET
Ideal 1959

Ideal equipped future fighter jet pilots with this electronic trainer making an imaginary flight demonstration into a special effects fantasy. After throttling up the motors, an array of levers, lights and gauges kept avid aviators occupied. Although both the airspeed and bank turn indicator operate automatically, constant attention must be given to guide the joystick. Failure to follow the radar scope blip will cause the "off course light" to blink.

Now for target practice, select one of six aircraft images from a built-in projector. Another switch moves those images across the wall into rocket range. FIRE 1-2-3-4! Okay, so I missed with the darts. On the next pass, I'll use the machine gun. My thumb is hovering over the joystick firing button. The planes are moving along the wall. Rat-a-tat-a-tat! The nice thing about machine gun sound effects is, you aim, you fire, and who's to say?

SKY DIVER
Remco 1962

Hit the switch to rev your engines. At full thrust (fresh batteries) the motor whines to a deafening screech, so much so that if you're indoors and Mom is on the phone, this feature will get you immediately kicked out of the house. On the way, don't forget the battery drive Tow. This treaded tractor connects a rod to the 30 inch jet and positions it for takeoff. We're almost ready. The ejection knob is fully wound. We've scouted for trees or bushes that could snag the chute before landing. I guess it's time to give our co-pilot the in-flight heave ho. Hit the switch. Co-pilot is jettisoned! GERONIMO! Just then, your neighbor's dog comes barreling into the yard heading straight for the parachute. You know the dog. The one named after Turkish Taffy ... Bonomo.

FLYING FOX
Remco 1960

Get ready for takeoff! It's time for a test flight of the Flying Fox Airliner. Fire up left and right engines. As all four propellers spin into a blur, pull the dual steering wheels up and go full throttle. Away we go! This airliner perched atop an aircraft control panel responds to movement of the steering wheels.

As we bank for turns, control panel indicators move in conjunction. Another lever blinks wing tip lights. Now bring to level flight and set auto-pilot. Wait a minute. May Day! May Day! Steepen the pitch and lower the landing gear. We're coming in for an emergency! What is it? Lunch time!

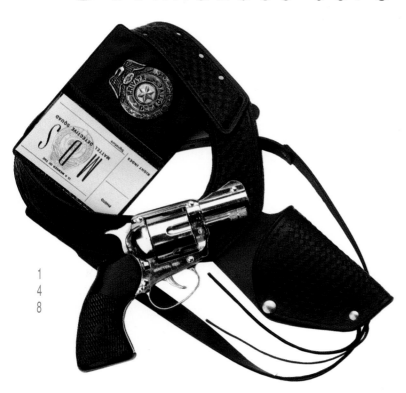

1
4
8

A local "classic rock" radio station annually celebrates one of television's grumpy cops on October 4th with "National Dan Matthew's Day". The reason being the numerical correlation between the date and a police radio sign off continually used by Broderick Crawford's *Highway Patrol* character, "10-4." This bit of trivia as well as the image of Dick Tracy in his yellow fedora, James Bond with a silencer gun, or hearing Sgt. Joe Friday's monotone interrogations permeates our very being. Faithfully we watched the devoted men and women who spied, car chased, dodged flying lead, and finally hand-cuffed crooks by story's end. Thanks to some terrific toy making, we too were able to carry a badge while upholding backyard and basement law and order.

SNUB NOSE 38 SHOOTIN' SHELL GUN SET
Mattel 1959

Become a Private Detective of the MDS (Mattel Detective Squad) with this Shootin' Shell .38 and Shoulder Holster set. The bullet firing cap gun also comes with a badge, identification card and wallet, Shootin' Shells, Greenie Stik-M Caps, and a pistol range target. Instructions recommend pasting your photo, pressing a fingerprint, and signing the ID card to give it the "real official look."

DICK TRACY COPMOBILE
Ideal 1963

A Dick Tracy Soaky bubble bath bottle checks out a most unique squad car called Copmobile. This is literally one toy to "shake a stick at" with a wooden hoop stick provided to chase this two foot long battery car around the floor. The idea is to direct the vehicle's movement by encircling the hood rod to steer left and right, or controlling the on/off, forward and reverse with the trunk rod. A hand held Kazoo speaker inserts for storage on a rear fender while the car's enormous roof speaker is a mystery I would like Mr. Tracy to investigate. It does absolutely nothing.

THE UNTOUCHABLES PLAY SET
Marx 1961

Cheese-it the Cops! It looks like Al Capone's heist is on hold with Elliott Ness and the Feds arriving on the scene. But wait! Rather than trying to make a quick getaway in the tin "Rolls," the scar-faced stogie-stuffed king of crime and his cronies may first want to heat up the streets. In order to play along, Marx issued a miniature cap gun for "pretend police" to shoot slugs at the thugs. Now keep your eyes peeled. It's not so easy to tell who's who. So far, we spotted a crook pulling a shotgun out of a golf bag and a suspicious violin case containing a Tommy gun. We also caught a glimpse of the Flapper who isn't really a weapon but whom some consider ... a knockout. Even though Marx's playset followed the *Untouchables* hit television series starring Robert Stack, few kids went Ga-Ga for Gangster shows resulting in stale sales. Today, the limited supply and hyped demand for this toy as a collectible has literally made it one of the Untouchables.

The Untouchables Play Set box appears courtesy of Chuck Rastelli
The Untouchables Play Set appears courtesy of Jeff Fisher

CRIME-BUSTER
Topper Toys 1965

From the makers of Johnny Seven O.M.A., comes its equivalent super action police gun. Crime-Buster is a multi-projectiled rifle which individually fires rifle bullets or sprays a four shot riot salvo. There is also a smoke grenade which comes with a bag of white powder for the smoking festivities. The most forbidding looking missile is really not even a weapon but instead used to send signals or messages. Other features and do-dads include a siren, adjustable stock, sparkling tracer barrel, and cartridge belt.

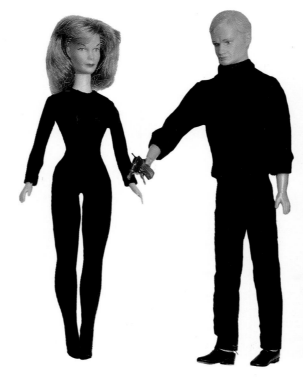

HONEY WEST
Gilbert 1965

ILLYA KURYAKIN
Gilbert 1965

Dressed in black getting ready for "The Blondes Have More Fun Affair" are action spy dolls of Honey West and The Man from Uncle's sidekick, Illya Kuryakin.

Honey West doll appears courtesy of Dick Brodeur

Illya Kurakin and Napoleon Solo dolls appear courtesy of Paul DeNero

DICK TRACY HAND PUPPET
Ideal 1961

DICK TRACY TWO-WAY WRIST RADIOS
Remco 1955

"Calling Dick Tracy. Calling Dick Tracy. Would you please stop pulling your wrist? You made me spill my milkshake." Such are the pitfalls of a wired pair of wrist radios. I'm sure Chester Gould's courageous crime stopper had the bugs worked out but we kids weren't always so lucky.

JOE JITSU HAND PUPPET
Ideal 1961

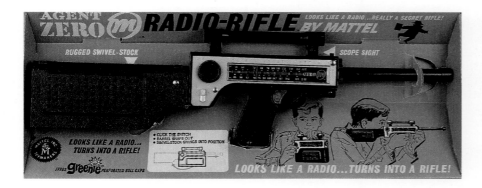

AGENT ZERO-M RADIO RIFLE
Mattel 1965

Mattel's Agent Zero-M disguises your means of defense with what appears to be a portable radio. When danger approaches, "turn on the radio" and out pops the barrel and stock of a cap firing rifle. Agent Maxwell Smart would be envious.

Agent Zero-M Radio Rifle appears courtesy of Harold Fischer

GUN OF THE GOLDEN AGENT
Hubley 1965

MULTI PISTOL 09
Topper Toys 1965

In the heyday of super spies and secret agents, Multi-pistol 09 was one of the many indispensable attaché cases that became standard issue to those on mysterious missions. If your duties required concealing hand guns that fire caps, shoot plastic bullets, an armor piercing rocket, torpedo, grenade, or secret message missile, Multi-Pistol filled the bill. If you need to go compact, a nifty little cap firing derringer slips out of the pistol's hollow stock for one last trick up your sleeve.

JAMES BOND CAP PISTOL WITH SILENCER
Lone Star 1964

The ultimately coolest capers in James Bond movies (not counting the kiss and kill love scenes) involved everyone sneaking around going Zip! Zip! Zip! with those silencer guns. This "precision" die cast cap gun comes with a snap-on plastic silencer and is marked James Bond 007 Automatic. It is appropriately made in England and promoted to be "From the film Goldfinger, now on release."

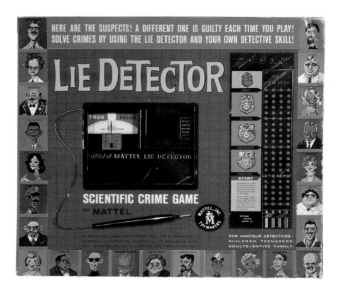

LIE DETECTOR
Mattel 1960

According to the School Teacher, "He had a mustache." The Car Salesman believes the suspect was "skinny." A clue from the Racketeer was that he had a "big jaw." Wait a minute. Racketeer? Car Salesman? Someone go out and get us an Official Lie Detector! Now, by inserting testimony cards to check the validity of each statement, evidence starts to lean towards certain suspects. The first person to correctly identify the criminal (of doing who knows what?) makes an arrest and wins. Points are accumulated to rise up through the force to Chief of Police.

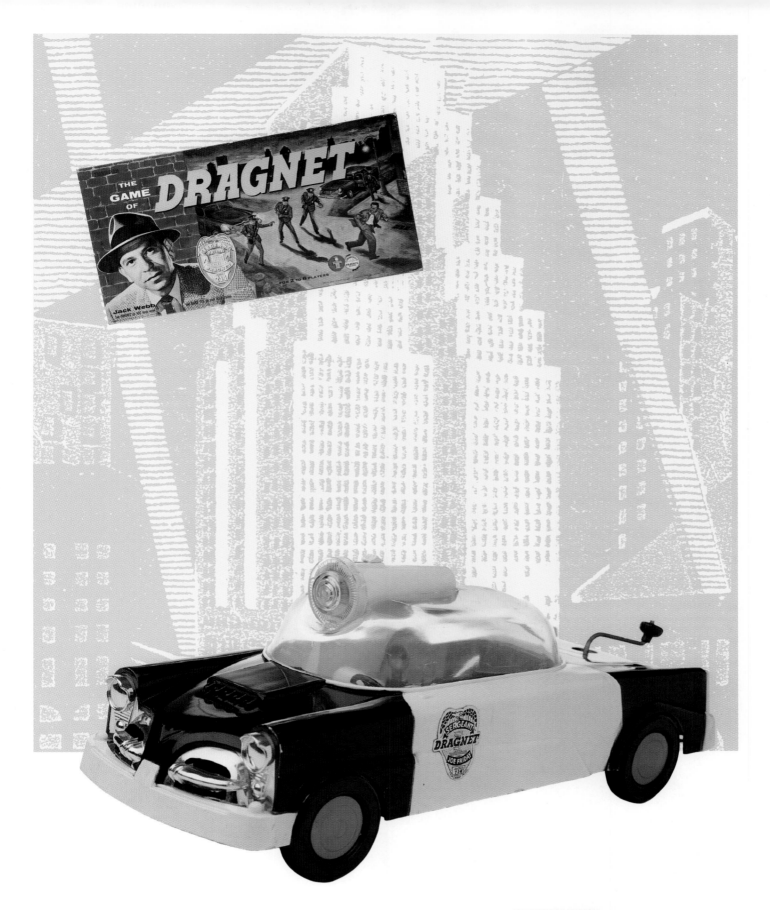

DRAGNET TALKING POLICE CAR
Ideal 1955

This is no mere squad car. It's a "mobile crime laboratory with everything you need to play *Dragnet*." Besides a flashlight mounted on the dome top, other accessories include a Tommy gun, riot gun, two police pistols and rifles, a camera, phone, and binoculars. The back seat area is set up with a map table and swivel chair, along with a gun case. But forget all that stuff. It's nothing compared to the hand crank record inside that begins with someone singing the theme's intro "Dum-de dum dum" then announces "Calling Sergeant Friday! Calling Sergeant Friday! Place dragnet for two escaped convicts ... that is all."

DRAGNET GAME
Transogram 1955

GOING GREAT GUNS

In *Dave Barry Turns 40*, the author humorist describes "Male Lifetime Phases," by classifying childhood into several age groups with associated interests. They are first, The Infancy Phase, age 0 to 2 with Interest Category : Pooping. Second, is the Innocence Phase, ages 3 to 9 with Interest Category : Guns. Then graduating from our youth to the Awareness Phase, ages 10 to 13 when the Interest Category becomes: Sex. As opposed to physical need, isn't it amazing that guns are singled out, as pre-occupying young minds with an intense desire to have and to hold.

FURY F-500
Nichols 1950's
    Am I dreaming or was there ever such a thing as a black and chrome battery operated cap firing Thompson 50 round machine gun embossed with outer space graphics? Somebody pinch me.

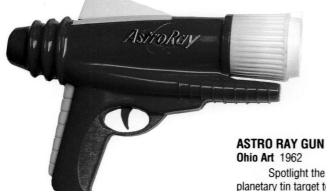

ASTRO RAY GUN
Ohio Art 1962

Spotlight the Astro Ray on an interplanetary tin target to score points blasting suction cupped play darts. For more space age fun the box charts Morse Code to send messages to other Astro friends.

TOM CORBETT ATOMIC RIFLE
Marx 1950's

All show and no go. This gorgeous looking ray gun spells doom for would-be attackers but not by shooting darts, balls, or making neat whizzing space noises. It's a clicker gun. That's right. An alien being is in your sights. You squeeze the trigger and ... Click!

ELECTRONIC SPACE GUN
Remco 1950's

For those who don't speak fluent box hype, we've provided a translation for Electronic Space gun's packaging promotion.

"Aim through the dual telescopic sights, pull the trigger, and see the Electronic light ray focus on the target." (It's a flashlight gun)

"The most thrilling toy of its kind." (At the time, the only one of its kind)

"The five color signal rays are selected by operating the automatic turret."(turn dial)

"Nuclear exhaust ports," "Color ray selector turret," and "Atom chamber" (Do-nothings)

"The high speed atom smasher whirls." (piece of cardboard with swirl design spins)

" ... generating atomic sound waves." (a boing noise)

PANTHER PISTOL
Hubley 1960

What in the wild wild west do we have here? "A flick of the wrist ... and it's in your fist!" Here's a 50 shot cap firing derringer that would have made James West proud. Only one drawback though ... wear short sleeves.

BLACK PIRATE FLINTLOCK
Hubley 1953

FLINTLOCK MIDGET
Hubley 1950's

In the 50's, an ornery kid with a double barreled flintlock could be transformed into Captain Kid, or with caps, "General Nuisance." As for Hubley's Midget, I remember these being given out at a friend's birthday party as favors. I guess it's time to re-access the past's insignificance.

PADLOCK PISTOL
Hubley 1961

Insert the key and give it a turn. True to its name, out pops a cap firing barrel. Gotcha!

COMMAND CANNON
Hasbro 1963

     Line up five incredibly mean looking pirates, ram down a cannon ball and take careful aim of this battery operated voice controlled cannon. With only five balls to shoot we can't afford a single miss. Now, give the command to ... Fire!

GUN THAT SHOOTS AROUND THE CORNER
Kenner 1964

     Is this what everyone needs? How many times have you laid on the floor behind the couch when a sibling sat down to read the comics and thought ... "If I only had that gun that shoots around the corner!" Aside from questioning the neccessity of this friendly weapon, who came up with the "catchy name" and who agreed with them?

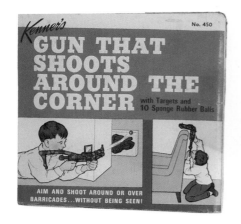

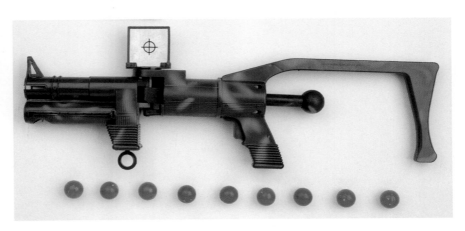

GATLING GUN
Rayline Toys 1960's

Without wheels, this version reminds me of the 3 Stooges' film where the monkey cranks a bullet belt through a meat grinder. Much the same way, the toy Gatling gun operates by clipping a magazine to the barrel, turn the crank and begin spraying 24 of your 100 bullets at little Yank and Reb figures. Incidentally, does anyone recognize these soldiers as blurry copies of Marx Civil War figures with reversed color schemes?

SCREAMING MEE MEE-E PISTOL
Remco 1965

The first kid on the block to show up with a Screaming Mee Mee-e was guaranteed to evoke a number of "Lemme Sees" from envious friends anxious to try it out. Coming from the spy era, Remco felt compelled to double up its firepower by also equipping the pistol to shoot caps. Why bother? It's already a scream!"

WESTERN GUN
Renwal 1950's

Here's one you won't lose in the grass ... a red, yellow, and blue bullet firing six shooter that comes in a cardboard holster.

MARK 'EM FLINTLOCK
Rayline Toys 1960's

Guess how Rayline's flintlock pistol fires bullets from a smoking barrel without caps or a flint wheel. Give up? First, take the toy powder horn and pour Mark 'Em powder into each bullet and seal. Ram the spring loaded barrel and you're ready to shoot. Now besides the smoking barrel, guess what happens upon impact? You Mark 'Em!

AIR BLASTER AND GORILLA TARGET
Whamo 1963

"The most amazing toy ever invented!"
"Shoot invisible air!" "One million free shots!"
"Shoot bugs, bounce balloons, burst bubbles,
fool your friends!" The box art graphic with
"fool your friends" suggests to sneak up
from behind and air blast your buddy's head.
So wouldn't "annoy your friends" be more
appropriate?" And did someone really con-
duct a ballistics test to see if you actually get
a million shots? I guess since mine still works
after 30 years they're probably in the clear.

FOOL YOUR FRIENDS

SHOOT BUGS

TRICK SHOT
Ideal 1961

"It shoots backward! It shoots forward!" View the trick shot
target in a mirror at the back end of the rifle, hold still, and squeeze the
trigger. Doggone if I didn't hit the target! It wasn't a bullseye but then it
probably wouldn't have been forwards either. A second trigger grip is
positioned for straight shooters.

It was torture! Sitting in the dark past midnight with the popcorn bowl on the floor down to a layer of burnt pieces and greasy unpopped kernels. Our *Chiller Theater* host, Chilly Billy, had promised the classics tonight, *Frankenstein, Wolfman,* and *The Mummy!* Keep in mind this pre-dates video taping and therewith speed search. So now, not only was I at the mercy of Hollywood's slow stingy style of waiting for glimpses of those beloved monsters but was also subjected to interruption by countless commercials selling used cars, Veg-A-Matics, and Garden Weasels.

My tiredness required closing eyes for longer and longer rest periods despite commands from the brain to stay awake. I started missing things. At one point there was Boris Karloff as Frankenstein leering into the gray screen but afterwards I must have dozed off totally missing the Wolfman. Finally, the Mummy's scream awakened me, but to my surprise, was coming from the hallway announcing "Time to get up!"

Now let us take a long look at some of those frightful toy favorites. The ones unleashed by day, and by night, hidden away in bedroom closets.

HAUNTED HOUSE GAME
Ideal 1962

Secret trap doors, ghosts, bats, and a vampire could pop out at any time. Then again they might not. It's all part of the mystery surrounding Ideal's Haunted House Game. Moves are determined by an owl spinner hooting as the number dial spins. The object is to be first in the attic trunk to nab the ruby treasure. Then make a quick getaway to reveal the hidden "You Win" sign.

WALKING FRANKENSTEIN AND MECHANICAL FRANKENSTEIN
Marx 1963

Once Boris Karloff portrayed Frankenstein for Universal Studios' motion picture, Mary Shelley's character had an unshakable image. In 1963, Louis Marx re-created those classic features with pale green skin, closed eyes, stitches, scars, and the neck bolts we've grown so fond of. Setting him in motion via wired remote control, mechanical clicking accompanies the stomp of each step. A looming tin litho body bends down to position huge hands for the grasping of unwilling objects.

Trying to make his getaway is the mechanized wind-up version. This little "Franky" was produced with a plastic head and metal legs, with either a plastic or tin torso. Finding any member of the Frankenstein family is a guaranteed thrill for fans of chills.

TALKING CASPER THE FRIENDLY GHOST
Mattel 1961

Like the cartoon character, Casper wants to make friends, "I'm a friendly ghost," "Will you play with me?" His ghostly body was originally white terry cloth but switched later to white fur possibly to keep him from being "C-O-L-L-L-D."

FRIGHT FACTORY
Mattel 1966

One of the favorites in Mattel's Thingmaker series, Fright Factory instructs kids to make the kind of spooky disguises they might hope to retrieve from a gumball machine. Only now you can stick them on your face! Then play witch doctor by gooping up a shrunken head that glows in the dark. Who could ask for anything more?

BATS IN YOUR BELFRY
Mattel 1964

How many people will answer yes to "Do you have Bats in your Belfry?" As promised, this game delivers flying bats sprung from a tower atop a creepy castle. Quick, grab your creature claw to see how many can be snatched in mid-air. Frightful fun for all.

UNIVERSAL STUDIO'S FRANKENSTEIN, MUMMY, WOLFMAN, AND CREATURE SOAKYS
IMCO 1960's

HERMAN MUNSTER HAND PUPPET
Ideal 1964

UNCLE FESTER HAND PUPPET
Ideal 1964

TALKING HERMAN MUNSTER HAND PUPPET
Mattel 1964

     The world's first friendly Frankenstein portrayed by Fred Gwynn tells kids, "I eat spinach for my complexion" and then invites them to a picnic " ... in the graveyard!" To put you at ease he compliments, "Oh you look nice ... just like I do," or expresses typical Munster optimism, "Cheer up. It's bound to get worse."

ADDAMS FAMILY GAME
Ideal 1964

MONSTER LAB
Ideal 1964

According to the toy box, Magilla Gorilla's cartoon program was interrupted to promote this light-up laboratory with what seems to be a nervous monster pacing back and forth. Once the controllers lead him to either end, he raises his arms, pops off his helmet to reveal a green skeletal face and shrieks.

Yeti and Whistling Spooky Kooky Tree appear courtesy of Arno and Anny Seeliger

YETI
Marx 1964

Perhaps better known as the Abominable Snowman, our white furry friend has a few tricks up his sleeve even for such an obvious monster. As expected he walks via wired remote control and raises or lowers those outstretched arms. Some may even be waiting for the mouthful of teeth to open and an obligatory scream. And scream he does with a loud screech which induces squinted eyes, winced expressions, covered ears, or a reflex response from parents, "Shut that thing off!" Kids loved it.

WHISTLING SPOOKY KOOKY TREE
Marx 1963

Is there such a thing as something so right ... it's wrong? This dark tree with crooked branches for arms and a bent stick nose scoots along the floor jerking its arms while alternately opening and closing eyes. As the name suggests an eerie whistle accompanies the enchanted movement. Fact is, it's scary. Too scary ... and sales were disastrous. Alterations were made such as lightening the bark color, adding white hands, nose, and more bright mushrooms around the tree's trunk. Still too scary ... just like those mean old trees in *The Wizard of Oz*. As toy legend has it, Marx ceased production within a year which accounts for the scarcity of this uniquely frightful toy.

HAMILTON'S INVADERS
Remco 1964

Creeping out of his cardboard cave, Horrible Hamilton and a giant bug buddy Spider crawl onto the scene, with antics activated by a pull string mechanism. Attacking with a Dwarf Tank, Mosquito Jeep, and Hornet Helicopter are an anonymous army known as the Blue Defenders who always hated and hunted these creatures. A Hamilton's Invaders Horrible Helmet and grenade launching cap gun supplied kids with the necessary hardware should they too decide to do something horrible.

Oddly enough, without character support from cartoons, comics, or TV, a monstrous toyline known as Hamilton's Invaders infested toy stores in the mid-60's. In vogue with ugly times they appealed to kids with a soft spot for Rat Finks, Nutty Mads, Weird Ohs, Foul-ups or space creatures from *The Outer Limits* known as the Zanti Misfits. Today, the few Invaders having survived extermination are being adored, sought, and bought by a cult following of collectors.

CAPTAIN ACTION
Ideal 1966

    I find the mystery of Captain Action not in his secret identity, nor for his popularity as a 1960's toyline but instead for the short lived success and disappearance of such a unique character. Looking back, it appears Ideal took steps to insure demand of their new hero. Starting with the idea of a posable action doll, changeable uniforms, and accessories Captain Action began by following the proven format of Hasbro's GI Joe. Next, Ideal chose various cartoon and comic costumes transforming the Captain and his Action Boy companion into various super heroes like Superman and Superboy, Aquaman and Aqualad, or Batman and Robin. The Captain himself went solo with a number of diverse impersonations ranging from The Lone Ranger, to Flash Gordon, and the Green Hornet. Even a deformed Dr.Evil came into the fold assuming a disguise to fool the good guys. Finally, Ideal capped it all off with the Silver Streak Amphibious Car, designed to fire missiles and send Captain Action and Action Boy on their next adventure.

Captain Action and CA Captain America appear courtesy of Chuck Eckles
CA Batman appears courtesy of Rick Rubis

Of all creatures, tall and small, be they wild beast or domesticated pets, most have been re-created into toy form at some time or another. Squeaking dogs, a growling lion, dinosaurs, a wacky bird, and a chimp on roller skates have each entered our homes to entertain and befriend. From playset figures to battery behemoths these colorful creatures came to life either mechanically or in our imaginations where they have remained ever since.

DANDY THE LION
Irwin 1963

PREHISTORIC TIMES PLAY SET
Marx 1961

In the 1950's and 60's Dinosaurs ruled the earth ... and every kid had some. Actually it was hard not to. A dollar would buy small herds on blister cards or in header bags. Others were given away free as cereal premiums or at gas stations. These plastic creatures were then taken on backyard safaris transforming the rocks, dirt, and tall grass into lost worlds while mud puddles and sandboxes became the tar pits. But! ... if you or a neighborhood buddy somehow lucked into getting a Marx Prehistoric Times play set, WOW! Now we're talking hard plastic rock formations, a lagoon, palm trees, ferns, and a cave! And where there are caves ... Cavemen! Hopefully, during play, you'll get an early draft choice of either a nifty Neanderthal or forbidding Dinosaur to use. On a bad day, you could get stuck with the guy making a fire or the skinny dinosaur that looks like a plucked chicken. Later after our growling and roars died off, we laid on the ground with the sand and dust of erupting volcanoes in our ears and underwear. Now it was time to mispronounce the impossible names stamped under each figure and compare their measurements to the length of cars, buses, houses, and ball fields. Or better yet, dream up acts of bravery should a Tyrannosaurus Rex come out of the woods that very moment!

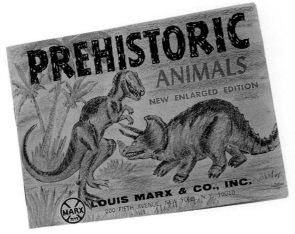

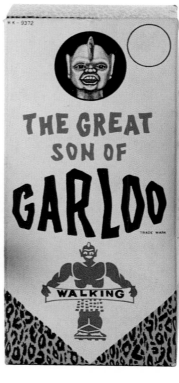

THE GREAT GARLOO
Marx 1961

For some, the stuff dreams are made of ... for others ... nightmares! Louis Marx's television toy commercial tries to carefully walk the fence appealing to both the wild and the mild with their promotion of the Great Garloo. First, they fantasize a typical monster movie in which Garloo looms around a mountain before wrecking havoc with a model railroad. Then inside the home, scary soundtrack is replaced with cheerful "Gumby music." Now the great gargantuan is gleefully carrying objects around the game room for brother while sister kisses him on the forehead! What is this? A girl kissing The Great Garloo? Are we selling toys or running for public office?

Besides our two foot tall mean green buddy, Marx produced a pair of wind-up "Sons." They differ only in that one's torso is cast in plastic like Dad's and wears a "Son of Garloo" medallion with metal chain. The other son's body is tin with the medallion being part of the lithographed design. At the flick of a switch, they whiz away in an arm and leg moving march.

Finally, two more Garloos were discovered in the Marx Toy Factory's research and development department at the company's 1976 liquidation sale. The plush and vinyl pair are believed by Marx Toy authority Gene Scala to be prototypes from the 60's that were never produced. They appear to be a Lady Garloo and Baby Garloo with magnetic discs to either hold each other's hands or for grasping small metal objects. Of the pair, it appears more time and effort was spent on Lady. Perhaps Baby was a last minute addition quickly assembled with a temporary generic face in time to make a "new toys proposal" meeting deadline. Reuniting them certainly creates a "strange but true" family portrait.

KING ZOR

KING ZOR
Ideal 1962

King Zor the Dinosaur is one cranky creature who, incidentally, does not want to play with you. He leers from alligator eyes, crawling lopsided along the floor growling the whole way. His jagged teeth and forked tongue quiver as if to warn ... Buzz off kid! So what does this forbidding image stir inside every goofy red blooded mischievous youngster? "Let's Get Him!" A dart gun was provided to do the dirty deed. The trick is to zap Zor in the circular target shaped tip of his tail. Once you've scored a direct hit, look out! He's mad! King Zor stops, backs up to face his assailant, fires a yellow ball from a hump back, snaps his jaw shut in rage, and heads straight for you! Now if this seems a bit violent rest assured. King Zor's designers promise an educational lesson for children playing with the toy, " ... if you hit something it will fight back."

ODD OGG
Ideal 1962

The commercial jingle goes "Odd Ogg, Odd Ogg, half turtle and half frog." Since most people haven't made the acquaintance of a half turtle half frog they are left wondering just what it is that an Ogg does. Again we turn to those sacred video vibes for the last line which informed us that "Odd Ogg plays ball." Okay, so he plays ball. What the jingle fails to inform is that this toy either has a unique sense of humor or is just plain ticklish. What happens is this. You are supposed to roll one of four balls underneath the toy. If your aim is true for the center, Odd Ogg politely scoots forward. But if the ball goes toward either side, he reels back razzing the attempt by squawking and flapping his mouth while sticking out a big red tongue. Although the directions credit a roll down the middle with scoring a point, guess where everyone aims?

CLANCY THE GREAT
Ideal 1964

Hats off to Clancy the Great. Toss in a coin and this colorful chimp turns your home into a roller rink. With a glance to the left and a glance to the right, he skates away adding an occasional squeak to his rhythmic repertoire. Now in case your last coin went into one of those toy candy banks, have Clancy wear his hat and simply press the button in his palm. Away we go again. One last feature which surprises a lot of people is that if you squeeze Clancy's head, it squeaks! Then again, don't most people make some sort of noise when you squeeze their heads?

SMARTY BIRD
Ideal 1964

The Smarty Bird (Smarticus Plasticae) is best known to frequent smooth hard surfaces like game room floors. In order to observe its behavior pattern, first try sneaking up from behind. Before it gets away, catch hold of the tail feathers and give a gentle turn to the right. Without a flutter, Smarty takes off in a beak flapping frenzy. Whistling while it works, the odd creature proceeds on rotating legs and circular webbed feet. The pupils of bulging eyes are also spinning in opposite directions as it scurries along. Now if, in your excitement, you accidentally turned the tail feathers to the left, jump back! Everything that happens is now happening in reverse. Should anyone be fortunate enough to track one of these down today, they are in for a rare tweet!

PENNY THE POODLE
Marx 1963

 Tagging along by a coiled cable, this pampered pup "springs" into action standing up, sitting, turning its head and "barking." A reversed color variation also produced Penny in pink.

LOBSTER RIDER
1965

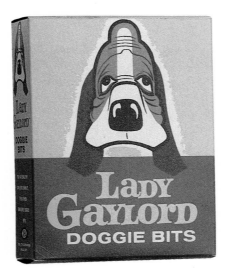

LADY GAYLORD
Ideal 1964

After Gaylord waddled his way into our homes, Ideal followed their popular pup with Lady Gaylord, a smaller pull toy version. Besides differing in methods of movement and size, Lady may be wearing her bonnet, or if not, can be distinguished by her painted eyelashes and hard plastic tail. Along with her dog house toy box, other terrific cardboard accessories included with Lady were a wheeled cart, box of Doggie bits, and food bowl.

GAYLORD
Ideal 1962

Gaylord the battery operated basset hound behaves much as the real thing, which is to say, with a very limited range of functions. At the tug of a leash, this plastic pedigree plods along taking slow high steps. Not high enough for a stairway, but maybe he could make it up a series of stacked books if so enticed with his toy bone waiting at the top. A metal strip wrapped around the center of the bone enables being picked up by the dog's magnetic snout. So, I guess you could say he has animal magnetism.

Once in a while, Gaylord's head dipping ear dragging strolls lead into a tight corner. Should this occur, another tug of the leash backs him out in reverse. This is, coincidentally, the same technique used by real-live basset hounds since they are not fond of bending at the middle.

Finally one last not-so-real function is a simulated barking sound made by squeezing a squeaker bulb in the leash handle. It's doubtful this squeak would scare anyone away but just to play it safe keep him quiet if you're expecting a toy delivery.

DANDY THE LION
Irwin 1963

Welcome to the Greatest Show in Toyland! Focus your attention to the center ring and meet the King of the Plastic Jungle ... Dandy Lion! This battery operated beast with emerald eyes locked in a perpetual stare stands poised for play. When ready for some ferocious fun, a flick of the switch sends the carnivorous creature crawling. A great growl is heard by all who behold his mechanical majesty come to life. Really there is nothing to fear since the pre-trained toy has already been taught its tricks. With Dandy, even the tiniest of tots can amaze their friends with lion taming skills. A "magic whip" was included to tap under his jaw which signals feline functions of forward, reverse, rearing up and sitting down. As with real lions, the instructions warn not to place fingers inside Dandy's mouth or to feed him anything. In case something would get swallowed, it is recommended to remove the animal's head. I bet if Dandy were to catch word of this, he would want a second opinion.

EPILOGUE:

And now for everyone who wonders what ever happened to their long lost toys, I have the answer. Stop blaming Mom. They all disappeared after that fateful night in 1964. The beginning of the end came from the same flickering box which first presented our terrific toys. The Ed Sullivan Show featured four young men who spoke the King's English but sang in an American dialect shaking their heads to "Yeah Yeah Yeah!"* Everyone dropped their toys and ran to the screen to see what all the clammer was about. We smiled, jumped up and down, and mimicked the shaking heads. On the next shopping trip, instead of looking for a toy, we used our green to buy black vinyl records which quickly turned to gold. Those "Pied Pipers" of rock and roll soon took all the children away from every village coast to coast never to return ... until today. May these images of Toyland's past take you back.

For readers who reached teenage in the 50's, substitute a pompadoured rockabilly star who curled his lip and got everyone "all shook up."

EXCUSES, EXCUSES

Writers love to state disclaimers. In my case it could be how impossible it is to present a complete volume of toys from any era. Guess what? At this time I would like to express deep regret for any toys omitted from these hallowed pages. In case your favorite toy is missing or was represented in an unfamiliar color variation, all I can say is ... Where were you during my vintage toy Plead - A-Thon? Actually, the standard excuse previously mentioned is true. Given the volume of toys mass produced from this time period being so vast, it would take a multi-volume set of toy encyclopedias to cover everything. Even department store Christmas catalogs fail to list all the toys available in one given year. I did try to basically represent the American toy market of two decades with a "something for everybody" approach emphasizing feature toys "As seen on TV!"

Addressing the task of dating our subject's vintage, a variety of sources were consulted such as department store catalogs, packaging, and the toy's markings. For the most part, these methods are extremely reliable however a few exceptions occasionally fool us. Another factor to keep in mind is that successful toys experience production runs which can last several years to an entire decade or longer without variation. My toy dating

system lists an exact year for the toy or particular variation shown according to its introduction into the toy market. When a date of origin could not be obtained, the entire decade encompassing its production was used as a time reference.

THE THANK PART

As it turned out, the calvary did show up in time. (Although my front yard will never be the same). I could have only achieved this undertaking with the help of an extremely cooperative army of people. I especially thank those who temporarily distanced themselves from some of their most prized possessions in order that we all could enjoy gazing at our old toys. I also want to take this opportunity to thank those who've had some monumental effect with this project or otherwise.

Dale Kelley of Antique Toy World Magazine was first to feature my articles under the heading Classic Plastic Toys. Dale has always been encouraging and remains one of few people who understands and is fascinated by the entire hobby beyond his own specific areas of interest.

Bill Manns of Zon International Publishing took a special interest in the plight of my "cheesy toys." Acting in the role of a mentor, Bill guided me through the maze of publishing into print. Without his expertise and devoted efforts, this endeavor would absolutely not have succeeded.

Jim Douglas used his tireless super-human abilities (focusing upside down in the dark) while taking on the task of capturing these classics despite their uncooperativeness (the potato heads kept falling over). During the entire session he maintained pride in each shot as if it were the one and only.

Ira Gallen from Video Resources of New York and host of Biograph Days Biograph Nights has made a career of restoring to screen, images of television's formative years. For this achievement I have bestowed the title, "Keeper of the White Dot." Our mutual obsession for television toys has enlisted us into a "secret squadron" devoted to the preservation of American Pop Culture which we proudly serve.

Dick Brodeur's award winning graphics, illustration, and an acute appreciation of 50's & 60's toys made him the perfect choice for this project. Throughout Toy Bop's conception he wore many hats to aid and abet its creation. In the beginning my request to Dick was simply ... "I'm not asking for much. Suppose at some future date the bomb goes off destroying the entire world and everything in it except two books. One should represent God

... the other should be something that represents the accomplishments, dreams, and fantasy of society ... Do you think we can do it?"

... And now Scarecrow, I think I'll miss you most of all ... er what I mean is ... Most of all I want to thank my wife Laurel for always believing in me.

Others in honorable mention include: Arno and Anny Seeliger, Les Petras, Francis Graham, Kathy Streten, Jared Bennett, Fred and Betty Limbach, Neil and Lois McElwee, Melanie and Kurt Pfaff, Gene Scala, Bertram Cohen, and Dick Beals. For extraordinary high honors ... Thanks Charlie, John, and Jim Frey, Mom, Dad and ... Santa Claus.

THE NIGHT BEFORE CHRISTMAS
Merrill Co. Publishers 1961

SOURCES

BILLY AND RUTH, AMERICA'S FAMOUS TOY CHILDREN 1951
TIME MAGAZINE December 12, 1955
PLAYTHINGS MAGAZINE December 1963
TOY & HOBBY WORLD March 1970
THE SATURDAY EVENING POST December 7, 1963
MARCH OF TOYS 1958
TOY HITS 1959
TOY FAIR 1959
MARX TOY FAIR PROMO 1959
THE TOY YEARBOOK 1960
SANTA'S OWN TOY BOOK 1961-63
TOYS AND GIFTS 1963
SPIEGEL 1963
MONTGOMERY WARD CHRISTMAS CATALOG 1953-54, 59-64, 66-68

PENNEYS TOY CATALOG 1963
SEARS CHRISTMAS CATALOG 1951-58, 63, 65, 68-70
HOWDY DOODY MERCHANDISE CATALOGUE (Kagran Corp. 1955)
TOPPER TOYS 1965
A TOY IS BORN by Marvin Kaye (Stein and Day 1973)
THE ROBOT BOOK by Robert Malone (Push Pin Press 1978)
INSIDE SANTA'S WORKSHOP by Richard C. Levy and Ronald O. Weingartner (Henry Holt 1990)
DAVE BARRY TURNS 40 by Dave Barry (Crown 1990)
TELEVISION TOYS, A VIDEO SERIES by Ira H. Gallen (Video Resources New York Inc. 1990)
HOW SWEET IT WAS by Arthur Shulman and Roger Youman (Bonanza 1966)
THE TOY INDUSTRY FACT BOOK (Toy Manufacturers of America 1993)

GOODNIGHT

GUMBY PUZZLE
Lakeside 1967

③ AFTER AIMING CANNON AT TARGET, CORRECT RANGE MAY BE FOUND BY RAISING OR LOWERING THE MUZZLE.

Wonderful Toys for a Merry Christmas

ATOMIC RAY

BIG LOO

Hubley golferino SCORE CARD T.M.

AND IS PUT TOGETHER AGAIN!

KING ZOR ™

MONSTER LAB $9.99 *without batteries*

Try to send the mechanical monster to opponent's end of lab by turning knobs. His direction is unpredictable

Monster Lab by IDEAL

COLUMN LEFT - MARCH!

REMOTE CONTROL BY SOUND

BLOWS APART UPON IMPACT!

$1⁶⁷

EMENEE

FOR **ACTIVE BOYS**